Contents

KV-579-557

GCSE
Foundation Maths 1

Caroline Paechter

Hutchinson
LONDON · SYDNEY · AUCKLAND · JOHANNESBURG

£4.25

Acknowledgements

Thanks are due to Ken Pollock for his ideas on the teaching of algebra and to Graham Millington for his help with the work on fractions. Thanks also to my fourth year pupils at Gladesmore Community School, Tottenham, whose comments and criticisms as they used the manuscript have been invaluable.

Hutchinson Education

An imprint of Century Hutchinson Ltd
62-65 Chandos Place, London WC2N 4NW

Century Hutchinson Australia Pty Ltd
89-91 Albion Street, Surry Hills,
New South Wales 2010, Australia

Century Hutchinson New Zealand Ltd
PO Box 40-086, Glenfield, Auckland 10, New Zealand

Century Hutchinson South Africa (Pty) Ltd
PO Box 337, Bergvlei, 2012 South Africa

First published 1985 as Arithmetic for You Book 1
Reprinted 1987, 1989

© Caroline Paechter 1985
Illustrations © Hutchinson Education

Designed, illustrated and photoset by ⟁ Tek Art Ltd., Croydon, Surrey
Printed in Great Britain by
Scotprint Ltd., Musselburgh

British Library Cataloguing in Publication Data

Paechter, Caroline
  GCSE Foundation Maths 1
  Bk 1
  1. Arithmetic-1961-
  I. Title
  513   QA107

ISBN 0 09 161271 3

Check your calculator work

In this book you will be able to use calculators for many sums. This section is to check how accurately you are using a calculator.

Use the answers in the back of the book to check each section as you go along.

Addition

A. **1.** 46 + 38   **3.** 318 + 177   **5.** 450 + 39   **7.** 128 + 612
     2. 102 + 49   **4.** 215 + 38   **6.** 612 + 48   **8.** 416 + 785

Subtraction

B. **1.** 312 − 56   **3.** 415 − 312   **5.** 520 − 17   **7.** 290 − 186
     2. 218 − 29   **4.** 654 − 426   **6.** 463 − 128   **8.** 473 − 412

Multiplication

C. **1.** 48 × 19   **3.** 54 × 5   **5.** 37 × 42   **7.** 27 × 43
     2. 27 × 16   **4.** 212 × 6   **6.** 55 × 36   **8.** 52 × 112

Division

D. **1.** 135 ÷ 3   **3.** 400 ÷ 16   **5.** 341 ÷ 22   **7.** 504 ÷ 21
     2. 1536 ÷ 32   **4.** 196 ÷ 14   **6.** 684 ÷ 19   **8.** 756 ÷ 18

Decimals

E. **1.** 4.5 + 2.9   **3.** 43.5 − 16.4   **5.** 2.6 × 3   **7.** 3.6 ÷ 1.2
     2. 6.3 + 17.2   **4.** 27.3 − 15.7   **6.** 4.8 × 2.1   **8.** 100.64 ÷ 3.4

Sharing it out

Sarah has a bar of chocolate.
She shares it with Paulette and Mustafa.
To be fair, she cuts it into *equal* shares.

These equal shares are thirds.
We write it: $\frac{1}{3}$ 1 share out of 3.

When we cut a whole one into equal shares we call the shares **fractions**.
Fractions get their names from the number of equal shares we cut the whole one into.

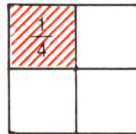

quarters	fifths	sixths	eighths
each one is $\frac{1}{4}$	each one is $\frac{1}{5}$	each one is $\frac{1}{6}$	each one is $\frac{1}{8}$
one share out of four	one share out of five	one share out of six	one share out of eight

Exercise 1

A. Copy the shapes below.
Share each one into the number of equal shares written inside it.
Name the fraction in words.
Shade one share. Name it using figures like the ones above.
The first one has been done for you.

1.

3 shares

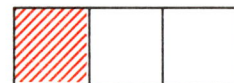

thirds
each one is $\frac{1}{3}$

one share out of three

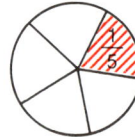

Now you do the rest in the same way.

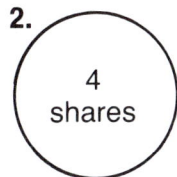

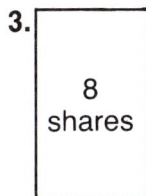

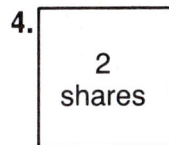

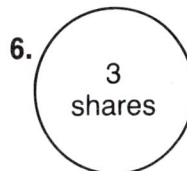

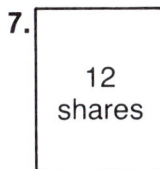

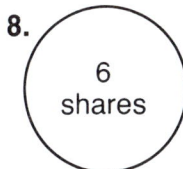

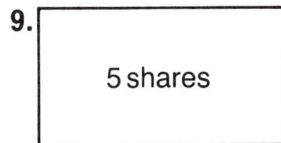

2. 4 shares

3. 8 shares

4. 2 shares

5. 10 shares

6. 3 shares

7. 12 shares

8. 6 shares

9. 5 shares

Draw some shapes of your own and share them.
Name the shares like the ones above.

B. Look at the shapes below.
Write down how many shares each one has been cut into.
Name the shaded parts in words and figures.
The first one has been done for you.

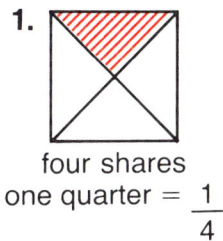

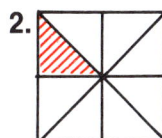

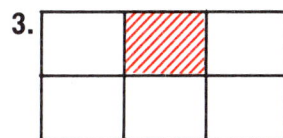

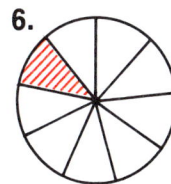

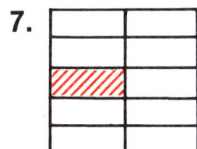

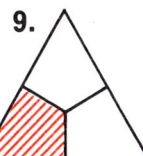

1.

four shares
one quarter = $\frac{1}{4}$

2.

3.

4.

5.

6.

7.

8.

9.

Thomas is greedy. When he shares his cake with Sally, he cuts it into three equal shares. Then he takes two shares for himself and gives one to Sally.

 ➡ 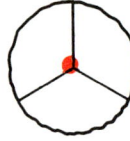 ➡

Thomas cuts the cake into thirds.

He takes two thirds
Sally gets one third.

We write two thirds as $\dfrac{2}{3}$

shows how many shares
the whole one is cut into ➡ $\dfrac{2}{3}$ ⬅ shows how many of
the shares we have

Exercise 2

A. Copy the shapes below.
Say in words and figures what fraction of each shape is shaded.
The first one is done for you.

1.

three quarters $\dfrac{3}{4}$

3 shares out of 4

2.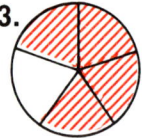

3.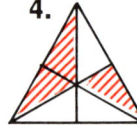

4.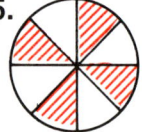

5.

B. 1. Draw a circle
Share it into sixths. Shade $\dfrac{5}{6}$

2. Draw a rectangle
Share it into eighths. Shade $\dfrac{7}{8}$

3. Draw a square
Share it into ninths. Shade $\dfrac{5}{9}$

4. Draw a circle
Share it into quarters. Shade $\dfrac{2}{4}$

5. Draw a rectangle
Share it into fifths. Shade $\dfrac{3}{5}$

8

Numbers in columns

We write numbers in columns to show how big or small they are.

MILLIONS	HUNDRED THOUSANDS	TEN THOUSANDS	THOUSANDS	HUNDREDS	TENS	ONES	
1	0	0	0	0	0	0	one million
	3	0	0	0	0	0	three hundred thousand
		2	0	0	0	0	twenty thousand
			4	0	0	0	four thousand
				3	0	0	three hundred
					6	0	sixty
						7	seven
3	0	5	0	0	4	8	three million, fifty thousand and forty-eight

We fill in any gaps on the right of the first digit with zeros. This is so that we can show how big a number is without writing in the columns.

Exercise 1

A. Draw a set of columns like the ones above.
Write the numbers below in the correct columns.
Don't forget to use zeros to fill in gaps to the right of the first digit.

1. six hundred
2. ninety-six
3. four thousand
4. fifty thousand
5. five hundred thousand
6. three million
7. thirteen thousand
8. two hundred thousand
9. one hundred and fifty thousand
10. nine million
11. five hundred thousand and forty-six
12. one hundred and eighteen
13. seven thousand five hundred
14. nine hundred thousand and ninety
15. six million five thousand
16. two thousand and one
17. fifty thousand and eighty
18. fifty thousand and eight
19. three million seven hundred thousand
20. forty-five thousand

B. Write these numbers in words.
Use the spelling list below if you need to.

MILLIONS	HUNDRED THOUSANDS	TEN THOUSANDS	THOUSANDS	HUNDREDS	TENS	ONES
6	0	0	0	0	0	0
	4	0	0	0	0	0
		3	0	0	0	0
			7	0	0	0
				6	0	0
					4	0
						8
7	4	0	0	0	0	0
		6	5	0	0	0
			9	9	5	0
		7	6	8	0	0
7	0	0	1	0	3	6
4	0	0	0	0	0	0
	3	5	0	0	0	0
	9	4	5	0	0	0
		1	5	0	0	0
			6	7	5	3
		3	4	9	8	7

Spelling list

1. one	11. eleven	30. thirty
2. two	12. twelve	40. forty
3. three	13. thirteen	50. fifty
4. four	14. fourteen	60. sixty
5. five	15. fifteen	70. seventy
6. six	16. sixteen	80. eighty
7. seven	17. seventeen	90. ninety
8. eight	18. eighteen	100. one hundred
9. nine	19. nineteen	1000. one thousand
10. ten	20. twenty	1 000 000. one million

When we add and subtract numbers we do not need to write in the column headings.
We still have to be careful to keep the numbers in the correct columns so we add or subtract the right digits.

Example

4023 + 762

Write the numbers one below the other in columns:

$$
\begin{array}{r}
4023 \\
+ 762 \\
\hline
4785 \\
\end{array}
$$

Then do the sum:

Exercise 2

A. Write these sums in columns.
   Then work out the answers.

1. 27 + 36
2. 386 − 244
3. 216 + 48
4. 1324 + 216
5. 41 + 218
6. 729 − 315
7. 416 − 84
8. 13 647 + 289
9. 657 − 49
10. 1034 − 963

11. 212 + 36 + 870
12. 310 + 481 + 639
13. 4137 − 986
14. 51 468 − 3487
15. 480 + 6371 + 48 895
16. 110 − 96
17. 430 − 289
18. 21 + 210 + 2100 + 21 000
19. 54 600 + 317 000
20. 4 730 000 − 2 581 000

B. Now do these.
   Write the numbers in columns one below the other.
   Then do the sum.

1. Three hundred and fifty add seventy-four.
2. Fifty-seven take away thirty-nine.
3. One thousand three hundred and forty take away nine hundred and twenty-two.
4. Fifty-five thousand four hundred add two hundred and eight.
5. Five million six hundred thousand add two hundred and thirty-five thousand.
6. Four hundred and nineteen take away two hundred and ninety.
7. Sixty take away seventeen.
8. Eighty-seven add seven hundred and sixteen.

More sharing

Shahid has a packet of biscuits. He shares them equally with Elaine and Gregory.

There are 30 biscuits in the packet.

10 + **10** + **10** = **30 biscuits.**

Shahid, Elaine and Gregory get 10 biscuits each.
Thirty shared into 3 equal piles gives you 10 in each pile.

So one third of 30 is 10. $\frac{1}{3}$ of 30 = 10.

Laura has £8. She shares it equally among her four children. Each child gets £2.

£2 + £2 + £2 + £2 = £8.

£8 shared between four people is £2 each.
So one quarter of £8 is £2.

$\frac{1}{4}$ of £8 = £2

Exercise 1

1. Jackie shares a box of 16 chocolates with Paulette.
How many does each girl get?
Copy and complete: one half of 16 is _____
$\frac{1}{2}$ of 16 = _____

2. Six friends share a box of 24 pens.
How many does each one get?
Copy and complete: one sixth of 24 is _____
$\frac{1}{6}$ of 24 = _____

3. Work these out. Write the answers the same way as before.
(a) 12 biscuits shared between 4 people.
(b) £10 shared among 5 people.
(c) 120 pencils shared between 3 people.
(d) £3.20 shared between 8 people.
(e) £4.80 shared between 4 people.
(f) 30 notebooks shared among 6 people
(g) £100 shared between 10 people.

Maxine is reading the school newspaper:
4G have done a survey.
They found that $\frac{3}{4}$ of the class hate school dinners.

How many pupils is this?

There are 24 pupils in 4G.

$\frac{1}{4}$ of 24 = 6

DAILY SKOOL
3/4 OF 4G
HATE
SCHOOL
DINNERS!

one quarter	one quarter	one quarter	one quarter
$\frac{1}{4}$	$\frac{1}{4}$	$\frac{1}{4}$	$\frac{1}{4}$

three quarters = $\frac{3}{4}$

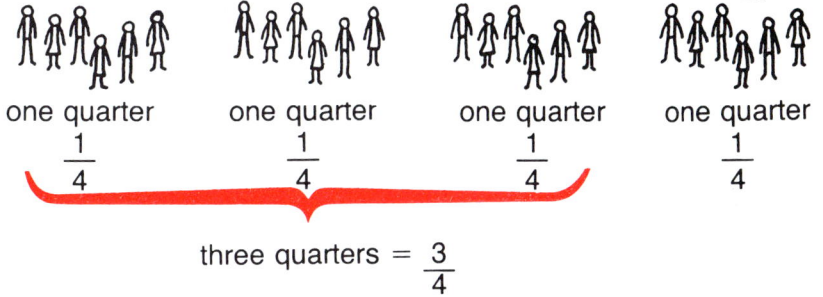

Three quarters of 24 is 3 groups of 6 = 18.

Exercise 2

1. There are 50 members in a youth club.
   One fifth like table football. How many is this?
   Four fifths like playing records. How many is this?

2. There are 33 people on a bus.
   One third are men. How many is this?
   Two thirds are women. How many is this?

3. There are 24 pupils in class 4B.
   One eighth have ginger hair. How many is this?
   Five eighths have dark brown hair. How many is this?

4. Paul has 18 biscuits.
   He gives one sixth of them to Tina. How many is this?
   He gives five sixths of them to Sabera. How many is this?

5. Samantha has £3.50.
   She gives one fifth of it to Rajinder. How much is this?
   She gives three fifths of it to Paulette. How much is this?
   How much does she have left? Write it as a share of £3.50.

6. Tom has £1.60.
   He gives one quarter of it to Paulo. How much is this?
   How much does he have left? Write it as a share of £1.60.

7. There are 48 biscuits in a tin.

   $\frac{1}{12}$ of them are shortbread. How many is this?

   $\frac{1}{8}$ of them are custard creams. How many is this?

   $\frac{5}{12}$ of them are chocolate covered. How many is this?

   The rest are ginger. How many is this?

8. 4G do a survey.

   They ask 60 people what pets they own.

   One third owned dogs. How many is this?

   One sixth owned cats. How many is this?

   One tenth owned goldfish. How many is this?

   Two fifths owned pigeons. How many is this?

9. Tricia has £7.20.

   She gives one eighth of it to Maxine. How much is this?

   She gives five eighths of it to Avril. How much is this?

   How much does she have left?

   Write it as a share of £7.20.

10. Darren has a box of 36 chocolates.

   He gives $\frac{1}{3}$ to Barjinder. How many is this?

   He gives $\frac{1}{4}$ to Michael. How many is this?

   He gives $\frac{1}{12}$ to Sharon. How many is this?

   He gives $\frac{2}{9}$ to Tricia. How many is this?

   How many does Darren have left?

   Write it as a share of 36.

Get the point

What's the point?

The decimal point.

10p is sometimes written as £0.10.
We can write this in columns in the way we did for whole numbers.

ONES	10ps	1ps
0 •	1	0

This means: no whole pounds
one 10p
no odd pennies

The **decimal point** shows where the pounds columns stop and the pence columns begin.

Exercise 1

Write these amounts of pence in columns like the ones above.
The first one has been done for you.

1. 25p   25p can be written £0.25

ONES	10ps	1ps
0 •	2	5

This means: no whole pounds
two 10ps
five odd pennies

Now you do the rest the same way.

2.	50p	**8.**	75p	
3.	30p	**9.**	15p	
4.	70p	**10.**	95p	
5.	35p	**11.**	11p	
6.	90p	**12.**	135p	
7.	40p	**13.**	250p	

Remember
When we write money with a '£' sign, we don't write a 'p' sign.

We can add and subtract money in the same way as ordinary whole numbers.
Make sure you keep the numbers in the correct columns.
Keep the decimal points underneath each other.

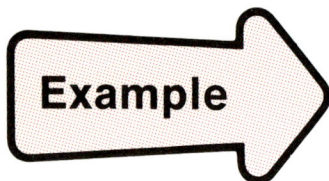

Example

27p + £1.36 + £2.40

Write the numbers
in columns:

```
    0.27
 +  1.36
    2.40
```

Then do the sum: £4.03   Total is £4.03

Exercise 2

Write these money sums in columns and work out the answers.

1. £1.35 + 55p
2. 27p + 95p
3. 45p + £3.50
4. £2.56 − 74p
5. £3.75 − £1.99

6. 49p + £1.27 + £0.35
7. 35p + £2.57 + £20
8. £3.70 − £1.95
9. £1.24 − 98p
10. £4.99 − £3.75

11. Maxine goes shopping. She has £3.50 in her purse. She buys a notebook costing 79p. How much does she have left?

12. Darren has £4.75. He buys a scarf costing £1.99 and a pen costing 45p. How much does he spend? How much does he have left?

13. Curtis has £2.95. He wants to buy a pair of trainers costing £7.50. How much more money does he need?

14. Tricia goes shopping. She starts off with £5.50 in her purse. She buys a bag costing £3.25 and a pair of gloves costing £1.99.
How much does she spend?
How much does she have left?
Tricia's friend pays her back £1.50 that he owes her.
How much does Tricia have now?

We have to be careful to keep the decimal point in the right place when we multiply and divide money by whole numbers.
Keep the numbers in columns.
Put the decimal points underneath each other.

Example

45p × 3

45p can be written as £0.45.

Write the numbers   0.45
in columns: ×   3
                     ─────
Then do the sum: 1.35   45p × 3 = £1.35

Example

£4.28 ÷ 4

Put the decimal points   1.07
underneath each other   4)4.28   £4.28 ÷ 4 = £1.07

Exercise 3

Now do these the same way:

1. £3.55 × 5   **6.** £9.72 ÷ 9
2. £1.80 × 4   **7.** 56p × 8
3. £2.75 ÷ 5   **8.** £1.36 ÷ 4
4. 75p × 4   **9.** £7.42 ÷ 7
5. £1.80 ÷ 6   **10.** £50.80 ÷ 2

11. Jackie buys 8 pairs of socks costing 65p each.
How much do they cost altogether?

12. Paul buys 3 shirts. They cost £3.75 each.
How much do they cost altogether?
How much change will he get from a £20 note?

13. Patricia buys 7 metres of cloth to make curtains. It costs her £16.45. How much did it cost per metre?

14. Eggs cost 96p per dozen. How much does one egg cost?

15. Which is cheaper, 2 packs of 3 cassettes costing £3.50 a pack, or 6 cassettes costing £1.25 each?

Latin and Greek

Many names for shapes come from Latin and Greek words. They tell us how many sides the shape has.

Many English words also come from Latin and Greek. We can use them to help us remember some shape names.

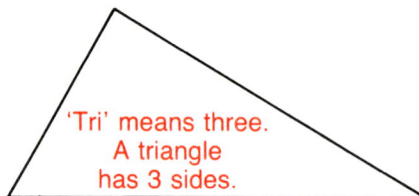

'Tri' means three.
A triangle
has 3 sides.

Other 'tri' words: trio, tricycle, triplets, tripod.

Exercise 1

Look at the shapes below.
Copy them into your book and answer the questions.

1.

'Quad' means four.
A quadrilateral
has 4 sides.

Find two more 'quad' words.

2.

'Pent' means five.
A pentagon
has 5 sides.

What 'pent' word names
an Olympic event?

3.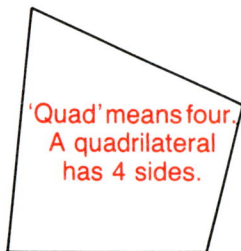

'Hex' means six.
A hexagon
has 6 sides.

4.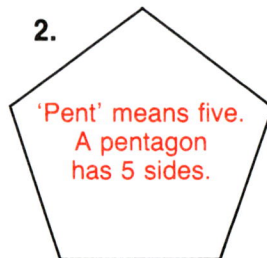

'Hept' means seven.
A heptagon
has 7 sides.

5.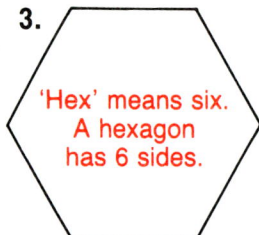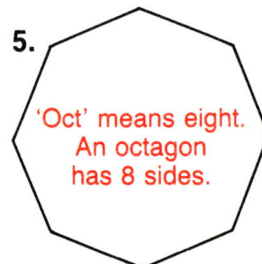

'Oct' means eight.
An octagon
has 8 sides.

What is the name for a
group of eight musicians?

6.

'Non' means nine.
A nonagon
has 9 sides.

7.

'Dec' means ten.
A decagon
has 10 sides.

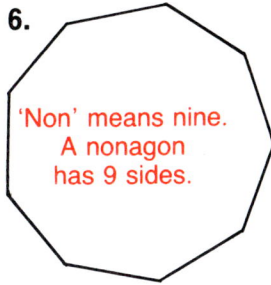

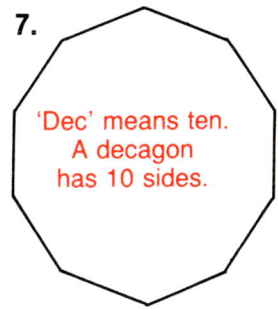

Find three more 'dec' words.

Now name these shapes by counting the sides.

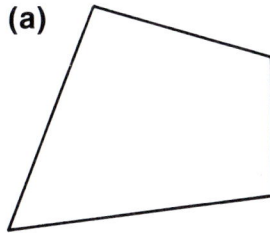

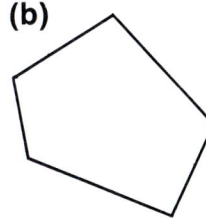

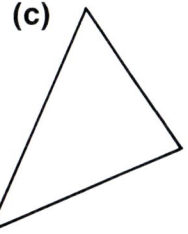

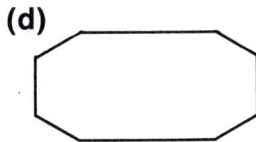

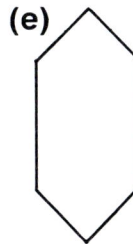

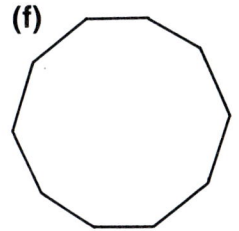

(a)

(b)

(c)

(d)

(e)

(f)

More Latin

'Sept' is another word for seven.
September used to be the seventh month.
What was the eighth month?
What was the ninth month?

'Cent' means 100.
There are 100 cents in a dollar.
Find five more 'cent' words.

'Bi' means two.
Bipeds walk on two legs.
Find as many 'bi' words as you can.

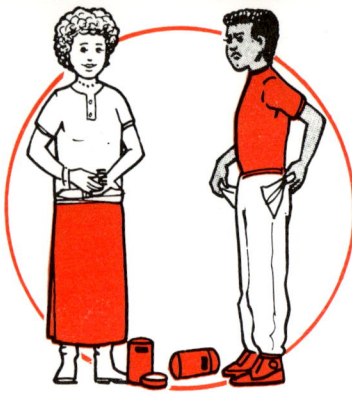

Is it enough?

Maxine wants to buy 4 pairs of ribbed tights.
They cost 95p a pair.
She has £4 in her purse. Is it enough?

95p is a bit less than £1.
So 4 × 95p is a bit less than 4 × £1.
 4 × 95p is a bit less than £4.
Maxine has enough money.

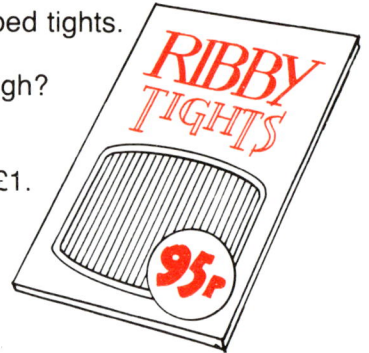

Curtis has 50p.
He wants to buy 5 pencils.
They cost 12p each. Has he got enough?

12p is a bit more than 10p.
So 5 × 12p is a bit more than 5 × 10p.
 5 × 12p is a bit more than 50p.
Curtis has not got enough money.

Exercise 1

1. Darren has £10. Is it enough to buy 2 shirts costing £3.95 each?

2. Alice has £5. Is it enough to buy 3 cassette tapes costing £1.95 each?

3. Ovid has 40p. Is it enough to buy 3 lb bananas costing 18p per lb?

4. Barjinder has £2.50. Is it enough to buy 5 notepads costing 45p each? Hint: 45p is a bit less than 50p.

5. Christina has £10. Is it enough to buy 6 gallons of petrol costing £1.85 a gallon?

6. Avril has £1.50. Is it enough to buy 3 coffee mugs costing 65p each?

7. Zahida has £25. Is it enough to buy 3 pairs of trousers costing £7.99 each?

8. Paulette wants to buy 2 pairs of gloves costing £2.85 each. She has £5. Is it enough?

9. Dermot wants to buy 3 pens costing 47p each. He has £1.50. Is it enough?

10. Ilknur wants to buy 2 pot plants costing £2.45 each. She has £5. Is it enough?

Example

Ovid is stocking up with pens for next term.
They cost 47p each.
He has £1.50. How many can he buy?

47p is a bit less than 50p.
There are 3 lots of 50p in £1.50.
So Ovid can buy 3 pens.

A teacher is buying calculators for school.
They cost £12.05 each.
She has £60. How many can she buy?

£12.05 is a bit more than £12.
There are 5 lots of £12 in £60.
But 5 lots of £12.05 is a bit more than £60.
So she can only buy 4 calculators.

Exercise 2

1. Gulsen is buying spider plants. They cost 97p each. She has £5. How many can she buy?

2. Tricia is buying goldfish for her little brother. They cost 55p each. She has £2. How many can she buy?

3. Dapinder is buying pencils. They cost 9p each. He has 70p. How many can he buy?

4. Ashmeed is buying postcards. They cost 11p each. He has £1.00. How many can he buy?

5. Joanne is buying apples. They cost 45p per pound. She has £1.50. How many pounds can she buy?

6. Sabera is buying videotapes. They cost £6.95 each. She has £30. How many can she buy?

7. Gurcan is buying folders for school. They cost 43p each. He has £2. How many can he buy?

8. Patricia is buying photo frames. They cost £2.05 each. She has £10. How many can she buy?

21

9. Terry is buying fish and chips for his friends. They cost £1.05 a portion. He has £6. How many portions can he buy?

10. Robert is buying batteries for his bike lights. They cost 55p a pair. He has £1.50. How many pairs can he buy?

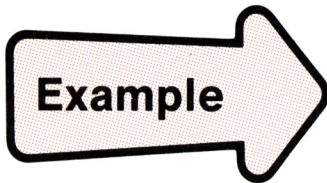

Example

Multiples

A **multiple** of a number is the answer you get when you multiply it by another number.

20 is a multiple of 4
because 5 × 4 = 20.
20 is also a multiple of 5.

Exercise 1

Find five multiples of each of these numbers:

4	10	3	5	8
25	50	100	250	1000

Check-up 1

1. Work these out using a calculator:
   (a) 36 + 15   **(b)** 43 − 29   **(c)** 47 × 3   **(d)** 108 ÷ 6

2. Draw a square.
   Share it into eighths. Shade $\frac{1}{8}$

3. Write this number in figures:
   one thousand two hundred and eight

4. Write this number in words: 35 056

5. There are 24 pupils in 4G.
   One eighth like reggae. How many is this?
   Three eighths like rock music. How many is this?

6. Work these out:
   (a)   £3.65 + 80p   **(b)**   £2.70 − £1.35

7. Tricia buys 3 skirts costing £4.75 each.
   How much do they cost altogether?

8. Name these shapes:

(a)   **(b)**   **(c)**

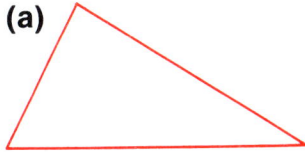

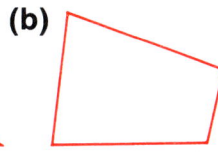

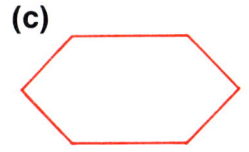

9. Ovid has £4.00. Is it enough to buy 5 T-shirts
   costing 85p each?

10. Patricia has £10. She wants to buy legwarmers costing
   £2.15 a pair. How many pairs can she buy?

11. Find three multiples of 15.

24 hours in a day

The **24-hour clock** is used for bus and train timetables. We use it instead of a.m. and p.m. to show morning and afternoon.

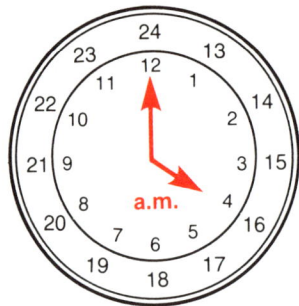

The times are the same until midday (12.00). Then we count on until 24.00 (midnight).

Sometimes midnight is written 00.00.

Remember: a.m. = morning
p.m. = afternoon

Example

24-hour clock times are always written using 4 figures.
1.00 a.m. is 01.00   1.00 p.m. is 13.00

Exercise 1

A. Using the 24-hour clock, how would you write:

1. 3.00 a.m.	**6.** 7.00 p.m.	**11.** 11.00 p.m.
2. 4.00 p.m.	**7.** 3.00 p.m.	**12.** 8.00 a.m.
3. midday	**8.** 5.00 a.m.	**13.** 4.00 a.m.
4. midnight	**9.** 6.00 p.m.	**14.** 9.00 p.m.
5. 1.00 a.m.	**10.** 2.00 p.m.	**15.** 10.00 p.m.

B. Write these as a.m. or p.m. times:

1. 05.00	**6.** 04.00	**11.** 13.00
2. 15.00	**7.** 06.00	**12.** 10.00
3. 12.00	**8.** 18.00	**13.** 17.00
4. 20.00	**9.** 22.00	**14.** 19.00
5. 01.00	**10.** 11.00	**15.** 08.00

C.
1. Maxine wants to catch a train that leaves at 16.00. When she arrives at the station her watch reads quarter past four. Will she catch the train?

2. Kevin's watch reads half past seven. He wants to catch a bus that leaves at 19.00. Is he in time?

3. Paulo wants to catch a train that leaves at 6.15 p.m. The station clock reads 18.00. Is he in time?

When we use the 24-hour clock we always show the minutes **after** the hours.

half past eight p.m. = 20.30
quarter to seven a.m. = 06.45

Remember: quarter **to** is 45 minutes **past** the previous hour.

Exercise 2

Using the 24-hour clock, how would you write:

1. 6.15 a.m.
2. 12.30 p.m.
3. 9.45 p.m.
4. 3.15 a.m.
5. 4.30 p.m.
6. half past 5 a.m.
7. quarter past 7 p.m.
8. quarter to 8 a.m.
9. half past 4 p.m.
10. quarter past 11 p.m.
11. quarter to 7 p.m.
12. quarter to midnight
13. half past 3 p.m.
14. quarter past 9 p.m.
15. quarter to 9 p.m.

When dealing with minutes **to** an hour, we must find the number of minutes **after** the previous hour.

We can make a sum to do this:

10 to 7 a.m.   60 minutes in 1 hour
                 so 10 to = 60 − 10
                          = 50 minutes **past**

so 10 to 7 a.m. is 06.50   50 minutes past 6.

With minutes **past** it is easy, e.g. 20 past 7 p.m. is 19.20.

Exercise 3

Write these as 24-hour clock times:

1. 10 to 8 a.m.
2. 20 past 9 p.m.
3. 20 to 1 a.m.
4. 10 past midday
5. 25 to 4 a.m.
6. 6.05 a.m.
7. 5 to 9 p.m.
8. 25 to 11 p.m.
9. 6.20 p.m.
10. 10 to midnight
11. 9.10 a.m.
12. 8.20 p.m.
13. 5 to 3 a.m.
14. 10 to 10 p.m.
15. 25 to 3 p.m.

Problems

We do not find sums on their own in real life.
We work out the answers to problems instead.
To work out the answer to a problem you need to know what sum to do.
Then you can do the sum.

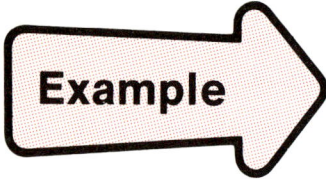

Example

Maxine has a packet of 30 biscuits.
She gives 16 to Roy and 5 to Tricia.
How many does she have left?

First we must work out how many biscuits Maxine gave away.
This is an addition sum:

$$\begin{array}{r} 16 \\ +\ 5 \\ \hline 21 \end{array}$$

Then we do a subtraction to find out how many biscuits are left:

$$\begin{array}{r} 30 \\ -21 \\ \hline 9 \end{array}$$

Maxine has 9 biscuits left.

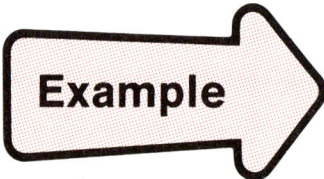

Example

Avril has a piece of string 96 cm long.
She cuts it into 3 equal pieces.
How long is each piece?

Avril's string has been divided into 3 pieces.
So we do a division sum:

$$3\overline{)96}^{\,32}$$

Each piece is 32 cm long.

26

Exercise 1

Find the answers to these problems.
Work out what sort of sum to do.
Then do the sum.

1. Roy has £35.
He buys a shirt for £8 and a jumper for £11.
How much does he have left?

2. Patricia has £20.
She shares it out between 4 people.
How much does each one get?

3. A teacher orders chalk in boxes.
Each box holds 10 sticks of chalk.
The teacher orders 8 boxes.
How many sticks of chalk does she have?

4. Maxine has £1.35 in her purse.
Her father gives her £2.50 and her grandmother
gives her 75p.
How much does she have now?

5. Pens cost 12p each.
How much will 9 pens cost?

6. A box of 5 cakes costs 70p.
How much do the cakes cost each?

7. In a car park there are 6 rows of cars.
Each row has 8 cars in it.
How many cars are there altogether?

8. Roy, Tricia and Sharon are playing cards.
Roy deals the whole pack of 52 cards.
How many cards does each one get?
How many cards are left over?

9. A swimming pool is 25 m long.
Renu swims 6 lengths.
How far does she swim altogether?

I have 100 legs

I have 1000 legs

Who needs more than 6?

Centimetres and millimetres

We use **centimetres** and **millimetres** for measuring many everyday things.

    centimetres = cm
    millimetres = mm

There are 10 millimetres in one centimetre
          10 mm = 1 cm

Look at the lines below.
They have been drawn on 2 mm graph paper to help you.
Each little square is 2 mm.
Five little squares is 1 cm.

———— 1 cm long
————————— 4 cm long
———————— 2 cm 4 mm long
——— 6 mm long

We usually use only one metric unit to write lengths.
We will write these lengths in cm.

1 mm = 0.1 cm
So we can write the lengths above as:
1.0 cm   4.0 cm   2.4 cm   0.6 cm

Exercise 1

Use the graph paper to measure the lines below.
Write the lengths in two ways.
The first one has been done for you.

1. ————————— 3 cm 4 mm
                    3.4 cm

Now you do the rest the same way.

2. —————————
3. —————————
4. ———————————————
5. ———
6. ——————————

7.
8.
9.
10.
11.
12.

Look at the line below.
It is 3.6 cm long.

We can write 3.6 cm in mm.
It is 36 mm long.   1 mm = 0.1 cm
3.6 cm = 36 mm.

Exercise 2

Use the graph paper to measure the lines below.
Write the lengths in cm and in mm.
The first one has been done for you.

1.   5.2 cm
                                      52 mm
2.
3.
4.
5.
6.
7.
8.
9.
10.
11.
12.
13.
14.
15.

29

Moving columns

A. Use a calculator to multiply these numbers by 10:

(a) 9   **(b)** 90   **(c)** 17   **(d)** 170   **(e)** 1700   **(f)** 875

What do you notice?

When we multiply a number by 10 it moves up one column to the left and we fill up the gap in the ones column with a zero.

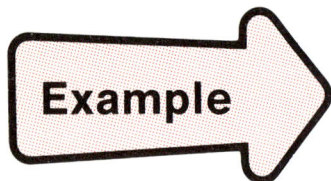

Example

17×10

	HUNDREDS	TENS	ONES
17		1	7
17×10	1	7	0

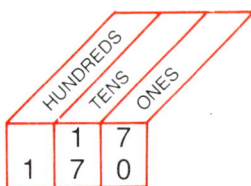

$17 \times 10 = 170$

17 moves one column to the left. We fill the gap in the ones column with a zero.

Exercise 1

Work out these multiplications by 10 without a calculator.

1. 12×10 **6.** 136×10 **11.** 1070×10
2. 5×10 **7.** 780×10 **12.** $313\,004 \times 10$
3. 15×10 **8.** 900×10 **13.** $420\,000 \times 10$
4. 24×10 **9.** 123×10 **14.** $27\,000 \times 10$
5. 39×10 **10.** 416×10 **15.** 999×10

When we divide a number by 10 it moves up one column to the right. We will need fewer zeros as there are now not so many columns to fill up.

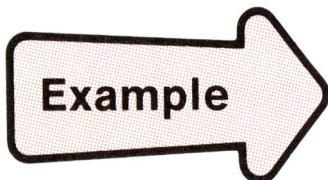

Example

$3500 \div 10$

	THOUSANDS	HUNDREDS	TENS	ONES
3500	3	5	0	0
$3500 \div 10$		3	5	0

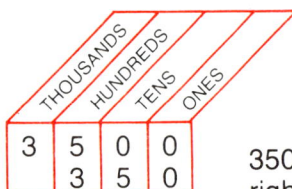

$3500 \div 10 = 350$

3500 moves one column to the right. We only need one zero to fill up the ones column.

Exercise 2

A. Work out these divisions by 10 without a calculator.

1. 350 ÷ 10	**6.** 780 ÷ 10	**11.** 1 000 000 ÷ 10
2. 420 ÷ 10	**7.** 7800 ÷ 10	**12.** 10 000 ÷ 10
3. 20 ÷ 10	**8.** 11 000 ÷ 10	**13.** 350 050 ÷ 10
4. 400 ÷ 10	**9.** 1340 ÷ 10	**14.** 43 000 ÷ 10
5. 5000 ÷ 10	**10.** 460 ÷ 10	**15.** 5 000 000 ÷ 10

B. Use a calculator to multiply these numbers by 100.

(a) 4   **(b)** 10   **(c)** 40   **(d)** 365   **(e)** 12

What do you notice?

Multiply the same numbers by 1000.
What do you notice?

When we multiply a number by 100 it moves up two columns to the left.
When we multiply a number by 1000 it moves up three columns to the left.
We fill up the gaps with zeros as before.

Example

45×100

45
45×100

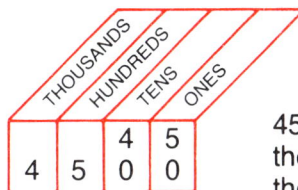

THOUSANDS	HUNDREDS	TENS	ONES
		4	5
4	5	0	0

$45 \times 100 = 4500$

45 moves up two columns to the left. We fill up the gaps in the tens and ones columns with zeros.

Example

9×1000

9
9×1000

THOUSANDS	HUNDREDS	TENS	ONES
			9
9	0	0	0

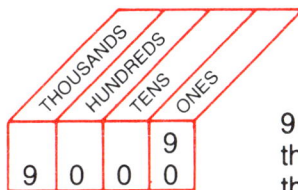

$9 \times 1000 = 9000$

9 moves up three columns to the left. We fill up the gaps in the hundreds, tens and ones columns with zeros.

Exercise 3

Work out these multiplications by 100 and 1000 without a calculator.

1. 5 × 100
2. 32 × 100
3. 45 × 1000
4. 76 × 1000
5. 999 × 1000
6. 780 × 1000
7. 19 × 1000
8. 120 × 100
9. 430 × 100
10. 12 × 1000
11. 3 × 1000
12. 30 × 100
13. 290 × 100
14. 29 × 1000
15. 350 × 1000

When we divide a number by 100 it moves up two columns to the right.
When we divide a number by 1000 it moves up three columns to the right.
As before we will not need so many zeros.

Example

3400 ÷ 100

THOUSANDS	HUNDREDS	TENS	ONES
3	4	0	0
		3	4

3400
3400 ÷ 100

3400 moves two columns to the right. We do not need any zeros as there are no gaps.

3400 ÷ 100 = 34

Exercise 4

Work out these divisions by 100 and 1000 without a calculator.

1. 100 ÷ 100
2. 3500 ÷ 100
3. 45 000 ÷ 1000
4. 8700 ÷ 100
5. 40 000 ÷ 1000
6. 40 000 ÷ 100
7. 67 000 ÷ 100
8. 7 000 000 ÷ 1000
9. 7 000 000 ÷ 100

Summary

To multiply a number by 10, move it one column to the left.
This is the same as adding one zero.
To multiply a number by 100, move it two columns to the left.
This is the same as adding two zeros.
To multiply a number by 1000, move it three columns to the left.
This is the same as adding three zeros.

To divide a number by 10, move it one column to the right.
This is the same as taking off one zero.
To divide a number by 100, move it two columns to the right.
This is the same as taking off two zeros.
To divide a number by 1000, move it three columns to the right.
This is the same as taking off three zeros.

Exercise 5

Work out the answers to these:

1. Gill wants to cover 10 books with paper. Each book needs a piece of paper 36 cm long. How many cm does she need altogether?

2. Paulette puts 100 matchsticks end to end. Each matchstick is 40 mm long. How far do they reach?

3. Jake has a length of wood 120 cm long. He cuts it into 10 pieces. How long is each piece?

4. There are 1000 m in 1 km. A stretch of motorway is 45 000 m long. How many km is this?

5. A tree grows 4 inches each year.
   How many inches will it have grown in 10 years?
   How many inches will it have grown in 100 years?

6. Barjinder is laying bricks. He can lay 175 bricks an hour. How many can he lay in 100 hours?

7. Darren is putting up a fence. It is 3500 feet long. He wants to put in a post every 100 feet. How many posts will he need?

Directions

Look at the map below.

I am in the city centre.

To get to Miles Hill I must go north.

To get to Osmondthorpe I must go east.

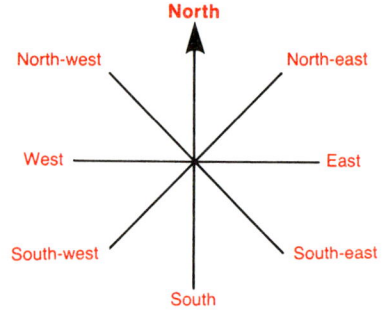

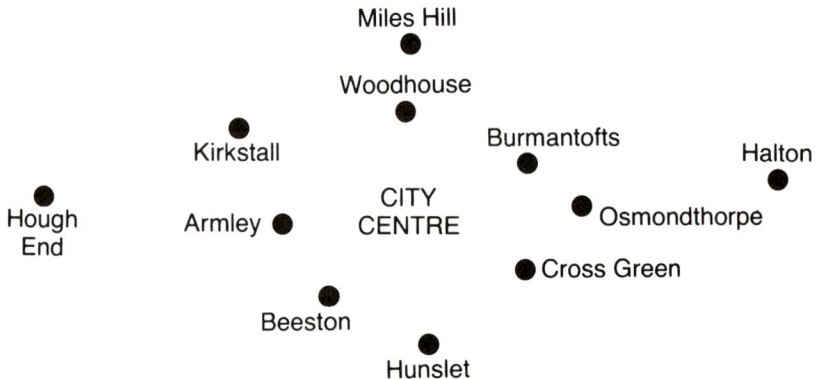

Miles Hill

Woodhouse

Kirkstall

Burmantofts

Halton

Hough End

Armley

CITY CENTRE

Osmondthorpe

Cross Green

Beeston

Hunslet

Exercise 1

1. I am in the city centre.
What direction must I go to get to:

  (a) Halton?   **(c)** Beeston?   **(e)** Cross Green?
  (b) Armley?   **(d)** Hunslet?   **(f)** Woodhouse?

2. I am at Osmondthorpe.
What direction must I go to get to:

  (a) City Centre? **(c)** Cross Green? **(e)** Burmantofts?
  (b) Halton?   **(d)** Hough End?   **(f)** Hunslet?

3. I am at Woodhouse.
What direction must I go to get to:

  (a) Miles Hill? **(c)** City Centre? **(e)** Kirkstall?
  (b) Burmantofts? **(d)** Armley?   **(f)** Hunslet?

Look at the map below.

The school is north of the shops.
The bridge is north-west of the church.

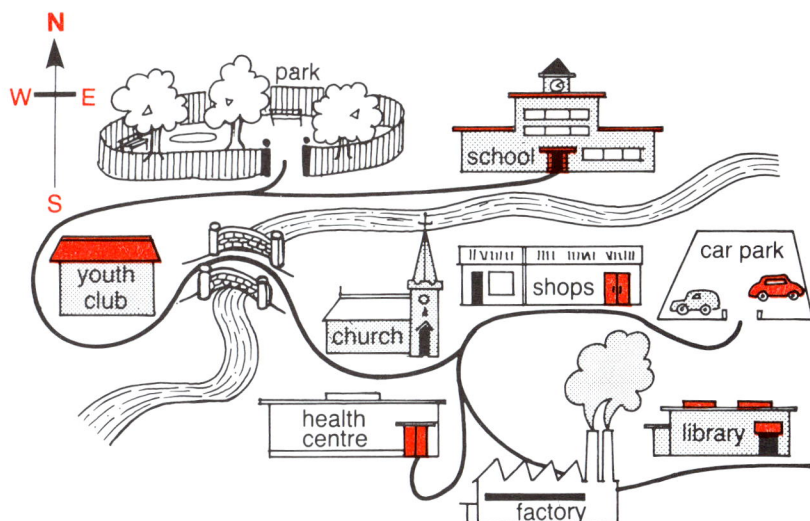

Exercise 2

1. What is north of the bridge?
2. What is east of the youth club?
3. What is south of the church?
4. What is north-west of the factory?
5. What is west of the library?
6. What is north of the health centre?
7. What is north-east of the health centre?
8. What is west of the car park?
9. What is north-east of the factory?
10. What is south of the school?
11. What is south-east of the park?
12. What is north-east of the youth club?
13. What is north-east of the church?
14. What is south-west of the park?
15. What is south-west of the shops?
16. What is west of the school?
17. What is south-east of the shops?
18. What is south of the park?

Maps are not always drawn with north pointing upwards. You have to look on the map for the arrow which shows north.

Look at the map below.
I am at the memorial.
To get to the bowling green I have to go north.

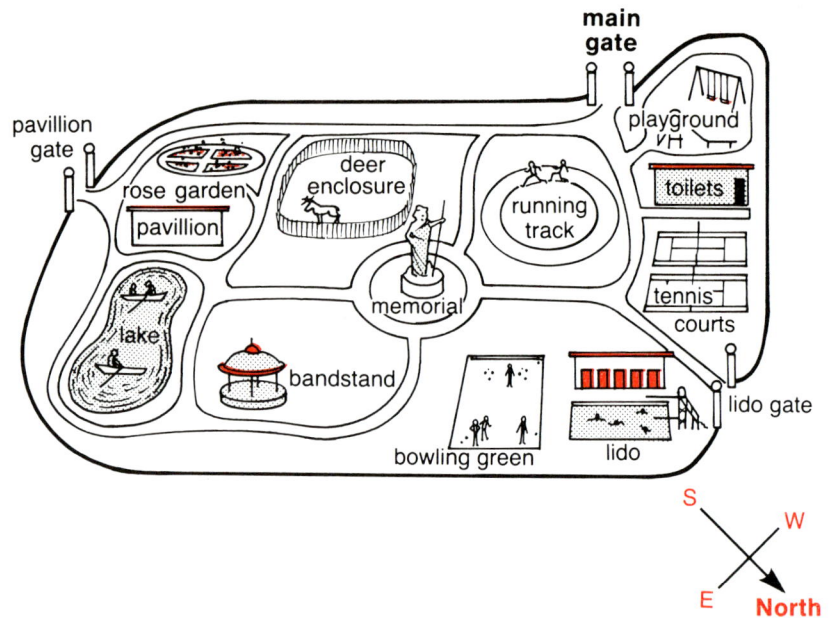

Exercise 3

1. I am at the memorial.
   What direction must I go to get to:

   (a) the deer enclosure?
   (b) the bandstand?
   (c) the main gate?
   (d) the lido?
   (e) the pavillion?

2. What is east of the playground?
3. What is south of the bandstand?
4. What is north of the running track?
5. What is south-east of the tennis courts?
6. What is east of the toilets?
7. What is west of the pavillion?
8. What is north-west of the memorial?

Calculators

In this book you will often be able to work out sums using a calculator.
That way, you save your brain for thinking.

We will use special symbols to show calculator keys.
They will be boxes with signs or numbers in them like this: $\boxed{\times}$ $\boxed{2}$

$\boxed{\text{c}}$ is a special sign

It stands for the button which clears the screen and leaves a zero like this:
$$\boxed{\qquad\qquad 0 \quad}$$

Always clear the screen before you start a calculation.
On calculators with a memory you may have to clear the memory, too.

Exercise 1

A. Rewrite the sums below without the calculator symbols.
Do them by pressing the keys in order.
The first one has been done for you.

1. $\boxed{\text{c}}$ $\boxed{3}$ $\boxed{\times}$ $\boxed{1}$ $\boxed{4}$ $\boxed{=}$ We write $3 \times 14 = 42$.

Now you do the rest the same way.

2. $\boxed{\text{c}}$ $\boxed{5}$ $\boxed{+}$ $\boxed{6}$ $\boxed{=}$ **6.** $\boxed{\text{c}}$ $\boxed{5}$ $\boxed{\times}$ $\boxed{8}$ $\boxed{0}$ $\boxed{=}$

3. $\boxed{\text{c}}$ $\boxed{6}$ $\boxed{+}$ $\boxed{5}$ $\boxed{8}$ $\boxed{=}$ **7.** $\boxed{\text{c}}$ $\boxed{9}$ $\boxed{\div}$ $\boxed{3}$ $\boxed{\times}$ $\boxed{1}$ $\boxed{8}$ $\boxed{=}$

4. $\boxed{\text{c}}$ $\boxed{8}$ $\boxed{\times}$ $\boxed{8}$ $\boxed{\div}$ $\boxed{4}$ $\boxed{=}$ **8.** $\boxed{\text{c}}$ $\boxed{4}$ $\boxed{+}$ $\boxed{3}$ $\boxed{-}$ $\boxed{2}$ $\boxed{=}$

5. $\boxed{\text{c}}$ $\boxed{3}$ $\boxed{2}$ $\boxed{-}$ $\boxed{1}$ $\boxed{6}$ $\boxed{=}$ **9.** $\boxed{\text{c}}$ $\boxed{5}$ $\boxed{3}$ $\boxed{-}$ $\boxed{2}$ $\boxed{8}$ $\boxed{=}$

B. Draw key symbols like the ones above to show how you would work these sums out on a calculator.

1. $3 + 6 + 17$ **3.** $6 \times 18 \times 30$ **5.** $298 - 14 + 63$
2. $9 \times 2 \div 3$ **4.** $51 + 17 + 134$ **6.** $1023 \div 3$

Columns for multiplying

Example

We need to keep numbers in columns when we multiply them.

47×3

Write it in columns:
```
     47
 ×    3
```
Then do the sum:
```
    141
```

Always multiply by the ones, then by the tens and so on.

Exercise 1

Write these multiplication sums out in columns and work them out.

1. 19×6	**6.** 44×3	**11.** 17×3
2. 24×5	**7.** 96×2	**12.** 75×6
3. 37×3	**8.** 54×4	**13.** 29×8
4. 54×9	**9.** 27×3	**14.** 37×4
5. 18×4	**10.** 14×8	**15.** 99×5

Example

30 is 3×10
So to multiply a number by 30 we can first multiply it by 3 and then multiply the answer by 10.

$15 \times 30 = 15 \times 3 \times 10$

Multiply 15 by 3:
```
     15
 ×    3
     45
```

Multiply the answer by 10: $45 \times 10 = 450$

$15 \times 30 = 450$

Example

500 is 5×100
So to multiply a number by 500 we can first multiply it by 5 and then multiply the answer by 100.

25×500

Multiply 25 by 5:
```
     25
 ×    5
    125
```

Multiply the answer by 100: $125 \times 100 = 12\,500$

$25 \times 500 = 12\,500$

3 Sardines × 5 = 15 Sardines

2 bones multiplied by 5 = 10 bones

Exercise 2

Work out these multiplications the same way.
The first one has been done for you.

1. 13×70

$70 = 7 \times 10$

Multiply 13 by 7:

$$\begin{array}{r} 13 \\ \times\ 7 \\ \hline 91 \end{array}$$

Multiply the answer by 10: $91 \times 10 = 910$

$13 \times 70 = 910$

Now you do the rest:

2. 15×20	**7.** 40×25	**12.** 64×400
3. 35×30	**8.** 80×15	**13.** 35×700
4. 9×50	**9.** 45×300	**14.** 500×27
5. 45×200	**10.** 95×70	**15.** 85×50
6. 5×500	**11.** 90×15	**16.** 66×300

17. A shopkeeper orders 50 boxes of tape cassettes. Each box contains 12 cassettes. How many cassettes does he order altogether?

18. In a multi-storey car park there are 6 levels. Each level can hold 70 cars. How many cars can the whole car park hold?

19. Pencils are sold in boxes of 12. A school buys 300 boxes. How many pencils do they buy altogether?

20. A gardener buys bulbs to plant in the park. They come in packets of 50. She buys 15 packets. How many bulbs does she buy?

21. Alice is building a wall. She needs 40 bricks for each row. The wall is going to be 13 rows high. How many bricks does she need?

22. A grocer sells nuts in 500 g bags. He has 25 bags. What is the total weight of the nuts?

23. Packets of crisps weigh 30 g. How much will a box of 125 packets weigh?

More than a whole one

Sometimes there is more than a whole one.
Five people want a quarter of a cake each.
But there are only four quarters in one whole one.
We need more than one cake.

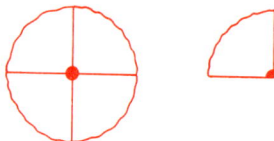

Five quarters is one whole one and one quarter.

We write five quarters $\frac{5}{4}$.

$$\frac{5}{4} = 1\frac{1}{4}$$

Exercise 1

1. How many whole ones are there in $\frac{3}{2}$? How many halves are left over?

Write it as $\frac{3}{2} =$

2. How many whole ones are there in $\frac{7}{4}$? How many quarters are left over?

Write it as $\frac{7}{4} =$

3. How many whole ones are there in $\frac{5}{3}$? How many thirds are left over?

Write it as $\frac{5}{3} =$

4. How many whole ones are there in $\frac{9}{5}$? How many fifths are left over?

Write it as $\frac{9}{5} =$

5. How many whole ones are there in $\frac{17}{8}$? How many eighths are left over?

Write it as $\frac{17}{8} =$

6. How many whole ones are there in $\frac{33}{10}$? How many tenths are left over?

Write it as $\frac{33}{10} =$

7. How many whole ones are there in $\frac{11}{4}$? How many quarters are left over?

Write it as $\frac{11}{4}$ =

Tricia works in a grocer's shop on Saturdays.
They sell monkey nuts in quarter-pound bags.
The manager asks her to weigh some nuts and put them into bags to be sold.

Tricia weighs the nuts. She has $1\frac{3}{4}$ lb.

How many bags does she need?

$1\frac{3}{4}$ lb is one whole pound and three quarters of a pound.

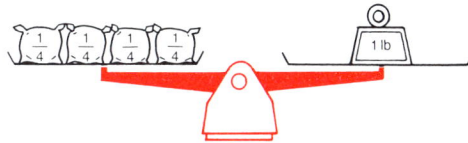

One whole pound is four quarters of a pound.
Tricia has another three quarters of a pound.
So altogether $1\frac{3}{4}$ lb will fill seven quarter-pound bags.

Exercise 2

1. On another day Tricia has $2\frac{1}{4}$ lb of nuts. How many quarter pounds is this?

2. John sells pieces of cake. Each cake is cut into 6 pieces. How many pieces are there in $1\frac{5}{6}$ cakes?

3. Gillian is building a wall. She has a space to fill that will take $3\frac{1}{2}$ bricks. But she only has half bricks. How many will she need?

4. Curtis's mum is the receptionist at the doctor's. The doctor sees one patient every quarter of an hour. How many patients can she see in $3\frac{3}{4}$ hours?

5. Ilknur goes swimming. She swims 3 lengths every minute. How many lengths can she swim in 4 minutes?

6. Roy sells biscuits in bags. Each bag weighs $\frac{1}{3}$ lb. How many bags will he need to hold $2\frac{2}{3}$ lb?

Parallel lines

Parallel lines never meet.
However long they are they never get any closer together.

Railway lines are parallel.
So are the wall bars in the gym.
So are the lines in your exercise book.

Exercise 1

Copy these drawings into your exercise book.
Colour lines parallel to each other the same colour.

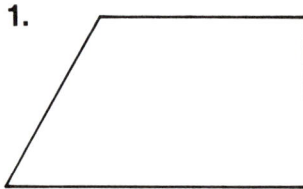

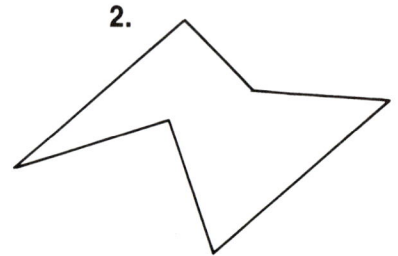

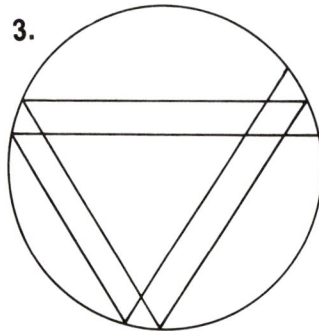

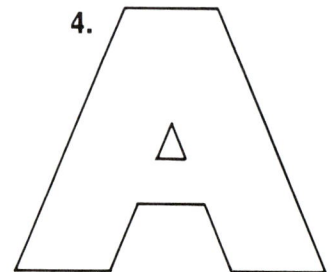

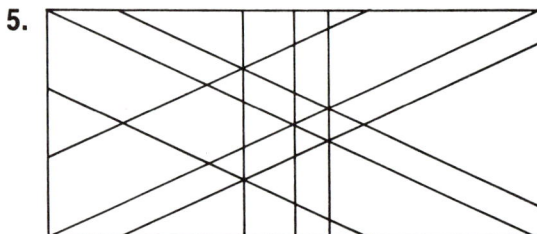

1.

2.

3.

4.

5.

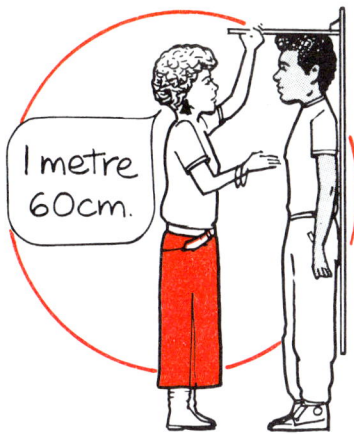

1 metre 60cm.

Metres and centimetres

We use **centimetres** for measuring fairly small things like pencils, books and tables.
For larger things, we can use **metres** (m).

'cent' means 'hundred'
e.g. 100 **cents** = 1 dollar
   100 years } = 1 **cent**ury
   100 years }

There are a hundred centimetres in one metre.

100 cm = 1 m

   1 cm = 0.01 m

Darren is 178 cm tall.
How tall is he in metres?

100 cm = 1 m
So Darren is 1 m 78 cm tall.
We can write this 1.78 m.

Darren's classroom is 4.8 m long.
How long is it in centimetres?

1 cm = 0.01 m
So the room is 480 cm long.

Exercise 1

A. Write these lengths in metres:

   1. A girl 164 cm tall.
   2. A room 384 cm long.
   3. A length of cloth 250 cm long.
   4. A bed 196 cm long.
   5. A cupboard 110 cm wide.

B. Write these lengths in centimetres:

   1. A room 2.65 m high.
   2. A length of rope 6.63 m long.
   3. A swimming pool 18.0 m long.
   4. A child 1.16 m tall.
   5. A pencil 0.12 m long.

Check-up 2

1. Change these to 24-hour clock times:
   (a) 8.00 a.m.   (b) 4.30 p.m.   (c) ten to 10 p.m.

2. Change these to a.m. and p.m. times:
   (a) 07.30   (b) 17.20   (c) 14.50

3. A mother shares £4.80 between her four children. How much does each one get?

4. Write these lengths in cm:
   (a) 1 cm 4 mm   (b) 3 cm 8 mm

5. Write these lengths in mm:
   (a) 2.3 cm   (b) 4.0 cm

6. Work out these:
   (a) 14×10 (b) 16×100 (c) $360 \div 10$

   (d) $4500 \div 100$

7. Look at the map on the right.

   (a) I am at Homely. What direction must I take to get to Turnbull?

   (b) Which town is:
       (i) North of Crickley?
       (ii) South-east of Tawley?

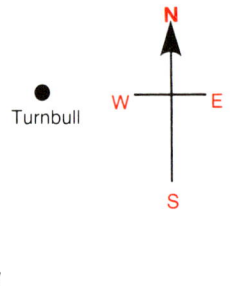

8. Work out these:
   (a) 18×4 (b) 15×30 (c) 17×200

9. How many whole ones in $\frac{19}{5}$? How many fifths left over?

   Write it as $\frac{19}{5} =$

10. How many quarter pound bags can you fill with $3\frac{3}{4}$ lb peanuts?

11. Write these in metres:
    (a) 312 cm   (b) 218 cm   (c) 46 cm

Factors and primes

The **factors** of a number are those numbers that can divide into it exactly with nothing left over.

5 is a factor of 15 because $15 \div 5 = 3$
$5 \times 3 = 15$

3 is also a factor of 15

18 is a factor of 36 because $36 \div 18 = 2$
$18 \times 2 = 36$

2 is also a factor of 36

1 is a factor of every number.
So is the number itself.

Example

Exercise 1

Find as many factors as you can of each of these numbers:

6,   15,   27,   20,   64,   16,   100,   81,   12

Prime numbers are numbers which only have themselves and 1 as factors.

29 is a prime number.
Its only factors are 29 and 1.

9 is not a prime number.
It has 3 as a factor as well as 9 and 1.

Exercise 2

Use a calculator to find out if these numbers are primes.
Remember, if you find any factors except the number itself and 1, the number is not prime.

11	18	29	106	87
123	19	47	1003	39
217	96	111	67	42

Shorthand

In **algebra** we often write shorthand.
We leave out some signs.

$4 \times a$ is written $4a$

$a \times b$ is written ab

$4 \times a \times b$ is written $4ab$

We leave out the $\times$ sign.
We always write the number first.

Exercise 1

A. Copy these and write them in shorthand.

1. $3 \times p$	**4.** $q \times 5$	**7.** $s \times t \times 3$
2. $4 \times b$	**5.** $6 \times r$	**8.** $a \times 35$
3. $p \times q$	**6.** $4 \times r \times s$	**9.** $a \times 2 \times b$

B. Copy these and put the signs back in.

1. $2a$	**4.** $13p$	**7.** $6pq$
2. $3b$	**5.** pr	**8.** $18rs$
3. $4r$	**6.** $8a$	**9.** $4abc$

Fractions are division sums.
In algebra, we write division sums like fractions.

$4 \div a$ is written $\dfrac{4}{a}$

$b \div 6$ is written $\dfrac{b}{6}$

$p \div r$ is written $\dfrac{p}{r}$

Exercise 2

Copy these and write them in shorthand.

1. $a \div 3$	**4.** $4 \div 3p$	**7.** $3p \div 5r$
2. $2 \div b$	**5.** $p \div q$	**8.** $pr \div ab$
3. $2a \div 3$	**6.** $2b \div p$	**9.** $ab \div pqr$

$2+2+2+2$ is 4×2 four lots of 2

In the same way $a+a+a+a$ is $4 \times a$ four lots of a

So we can write $a+a+a+a$ as $4a$ in shorthand.

Exercise 3

Copy these and write them in shorthand.

1. $a+a+a$ **4.** $p+p+p$ **7.** $l+l+l+l$

2. $b+b$ **5.** $a+a+a+a+a$ **8.** $c+c+c+c+c+c$

3. $p+p+p+p$ **6.** $r+r$ **9.** $d+d+d$

$$4a = a+a+a+a$$

$$2a = a+a$$

So $4a + 2a = (a+a+a+a) + (a+a)$

$$= 6a \qquad \text{six lots of a}$$

Exercise 4

Work out these additions in shorthand.

1. $2a+3a$ **4.** $7p+3p$ **7.** $10y+10y$

2. $4b+5b$ **5.** $2m+3m$ **8.** $3r+4r$

3. $6b+2b$ **6.** $4m+6m$ **9.** $5r+2r$

We can subtract using shorthand, too.

$3b - b = b+b+b-b$ b means $1 \times b$

$$= b+b$$

$3b - b = 2b$

Exercise 5

Work out these subtractions in shorthand.

1. $3p-2p$ **4.** $2m-m$ **7.** $7p-4p$

2. $4a-a$ **4.** $5m-2m$ **7.** $6m-2m$

3. $6a-3a$ **6.** $21r-14r$ **9.** $9y-y$

Division sums

$\frac{1}{3}$ is 0·3333 recurring

We can write division sums as fractions.
e.g. 3 ÷ 4 can be written as $\frac{3}{4}$

It means '3 whole ones share into 4 equal parts'.

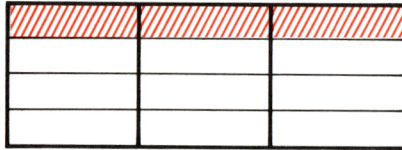

each part is three quarters of one whole one.

Work out 3 ÷ 4 on the calculator.
$$3 ÷ 4 = 0.75$$

We can say that $\frac{3}{4}$ is the same as 0.75

Fractions are division sums.
To change a fraction to a decimal, we do the division sum.

Exercise 1

Use the calculator to change these fractions to decimals.

1. $\frac{1}{4}$ **4.** $\frac{1}{5}$ **7.** $\frac{3}{8}$ **10.** $\frac{5}{8}$

2. $\frac{1}{2}$ **5.** $\frac{4}{5}$ **8.** $\frac{1}{3}$ **11.** $\frac{1}{16}$

3. $\frac{1}{10}$ **6.** $\frac{7}{10}$ **9.** $\frac{2}{3}$ **12.** $\frac{1}{9}$

You will have noticed that some fractions give decimal numbers that repeat.

For example, $\frac{1}{3} = 0.3333 \ldots$

This is because 1 cannot be divided **exactly** by 3 using decimals.
There is always a little bit left over.
We call these numbers '**recurring decimals**'.
We show this with a little dash (') after the first two numbers

e.g. $\frac{1}{3} = 0.33'$

48

Sometimes we need to change whole numbers and fractions into decimals.

We just change the fraction.

The whole number stays the same.

Change $2\frac{4}{5}$ to a decimal.

The 2 stays the same.
We do a division sum to change $\frac{4}{5}$ to a decimal.

$4 \div 5 = 0.8$

So the answer is $2 + 0.8 = 2.8$

$2\frac{4}{5} = 2.8$

Exercise 2

Use the calculator to change these to decimals:

1. $1\frac{1}{2}$

2. $2\frac{1}{4}$

3. $3\frac{2}{5}$

4. $1\frac{3}{4}$

5. $6\frac{3}{8}$

6. $2\frac{1}{10}$

7. $4\frac{1}{3}$

8. $2\frac{5}{8}$

9. $3\frac{7}{10}$

10. $4\frac{3}{5}$

? Special quadrilaterals

Quadrilaterals are shapes with four sides and four angles.
Special quadrilaterals have special names.
The names describe the quadrilaterals.

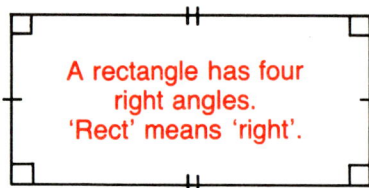

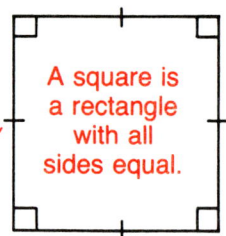

A rectangle has four
right angles.
'Rect' means 'right'.

these marks
show which
sides are equal

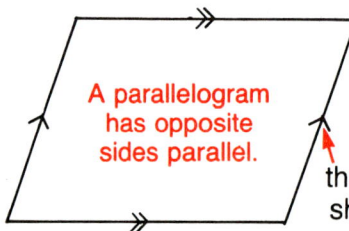

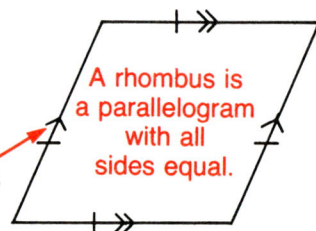

A square is
a rectangle
with all
sides equal.

A parallelogram
has opposite
sides parallel.

these marks
show which
sides are
parallel

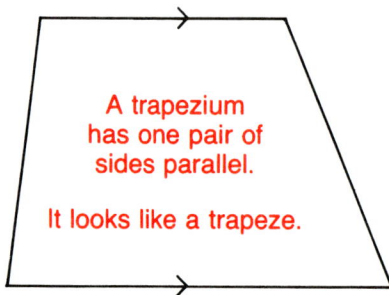

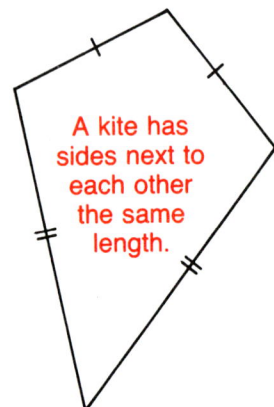

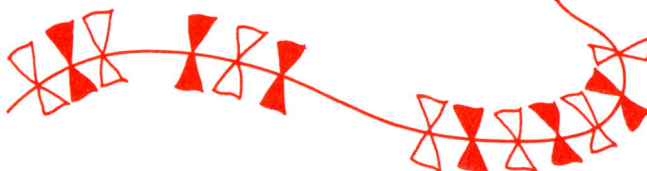

A rhombus is
a parallelogram
with all
sides equal.

A trapezium
has one pair of
sides parallel.

It looks like a trapeze.

A kite has
sides next to
each other
the same
length.

Exercise 1

Name the shapes below.
Use the drawings on page 50 to help you.

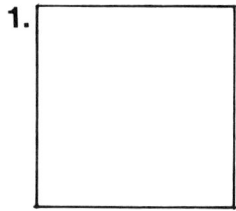

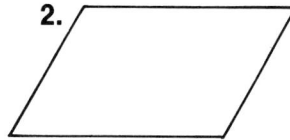

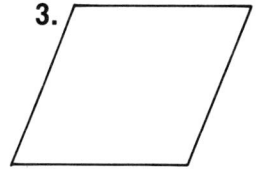

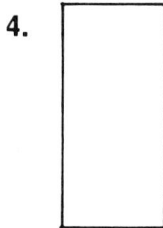

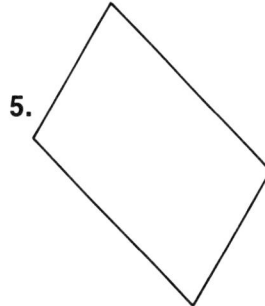

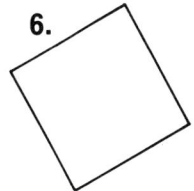

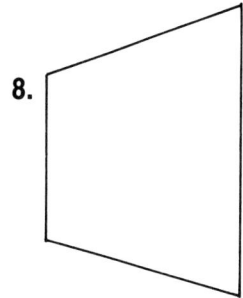

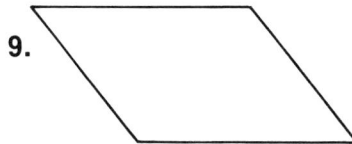

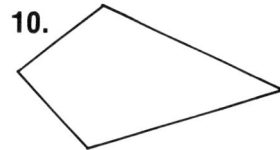

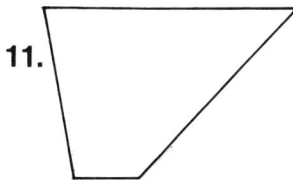

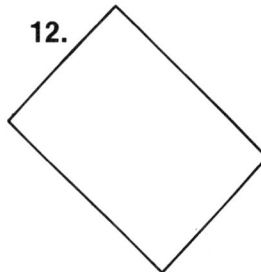

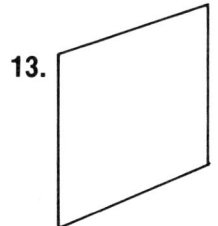

1.

2.

3.

4.

5.

6.

7.

8.

9.

10.

11.

12.

13.

The same thing

You will need cm² paper, pencil, ruler, scissors.

Draw three squares 6 cm long, 6 cm wide.
Share them like the ones below.

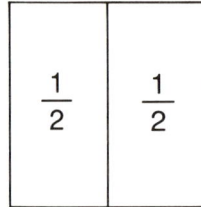

2 shares

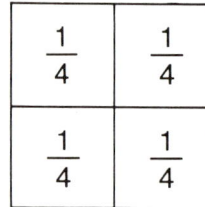

4 shares

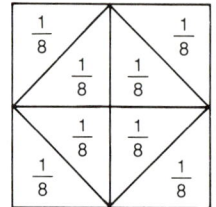

8 shares

Cut out the squares. Cut them into the shares you have drawn.

How many quarter shares do you need to cover one half?

One half is the same as two quarters $\frac{1}{2} = \frac{2}{4}$

Exercise 1

Work these out in the same way. Write your answers out as we did above.

1. How many eighth shares do you need to cover one quarter?
2. How many eighth shares do you need to cover one half?
3. How many eighth shares do you need to cover three quarters?

   Take another 6 cm × 6 cm square. Cut it into sixteenths.

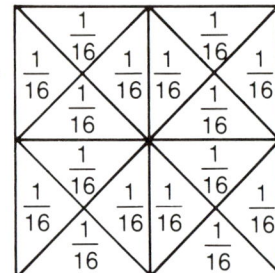

16 shares

4. How many sixteenths is the same as one half?

5. How many sixteenths is the same as one quarter?

6. How many sixteenths is the same as three quarters?
7. How many sixteenths is the same as one eighth?
8. How many sixteenths is the same as three eighths?
9. How many sixteenths is the same as five eighths?
10. How many sixteenths is the same as seven eighths?

Draw four more squares 6 cm × 6 cm.

Share them out like the ones below, then cut them out.

$\frac{1}{3}$			$\frac{1}{6}$	$\frac{1}{6}$		$\frac{1}{9}$	$\frac{1}{9}$	$\frac{1}{9}$		$\frac{1}{12}$	$\frac{1}{12}$	$\frac{1}{12}$	$\frac{1}{12}$
$\frac{1}{3}$			$\frac{1}{6}$	$\frac{1}{6}$		$\frac{1}{9}$	$\frac{1}{9}$	$\frac{1}{9}$		$\frac{1}{12}$	$\frac{1}{12}$	$\frac{1}{12}$	$\frac{1}{12}$
$\frac{1}{3}$			$\frac{1}{6}$	$\frac{1}{6}$		$\frac{1}{9}$	$\frac{1}{9}$	$\frac{1}{9}$		$\frac{1}{12}$	$\frac{1}{12}$	$\frac{1}{12}$	$\frac{1}{12}$

3 shares   6 shares   9 shares   12 shares

Exercise 2

Use your pieces to work these out. Write your answers in the same way as for Exercise 1.

1. How many sixths is the same as one third?
2. How many ninths is the same as one third?
3. How many sixths is the same as two thirds?
4. How many ninths is the same as two thirds?
5. How many ninths is the same as four sixths?
6. How many twelfths is the same as one third?
7. How many twelfths is the same as one sixth?
8. How many twelfths is the same as two thirds?
9. How many twelfths is the same as three ninths?
10. How many twelfths is the same as four sixths?
11. How many ninths is the same as two sixths?

Compare these pieces with the ones you cut out before.

12. How many sixths is the same as one half?
13. How many twelfths is the same as one half?
14. How many twelfths is the same as one quarter?

15. Work out how many tenths is the same as one fifth. Draw and share squares to check your answer.

We call fractions which are exactly the same size **equivalent fractions**.

Fractions are equivalent if they are the same amount of a whole one.

For example, we saw when we shared squares that $\frac{3}{9}$, $\frac{2}{6}$ and $\frac{1}{3}$ are all the same.

We say that $\frac{3}{9}$, $\frac{2}{6}$ and $\frac{1}{3}$ are equivalent.

We can work out equivalent fractions without drawing.

$$\overset{\times 2}{\frac{1}{4} = \frac{2}{8}}\underset{\times 2}{}$$

As long as we always multiply both the top and bottom numbers by the same amount, we will get equivalent fractions.

We can use this method to make sets of fractions which are equivalent.

e.g. $\quad \overset{\times 2 \; \times 3}{\frac{1}{2} = \frac{2}{4} = \frac{6}{12}}\underset{\times 2 \; \times 3}{}$ or $\overset{\times 3 \; \times 3}{\frac{1}{2} = \frac{3}{6} = \frac{9}{18}}\underset{\times 3 \; \times 3}{}$

Exercise 3

Use this method to make sets of fractions equivalent to the ones listed below.

Stop when the bottom number is about 30!

You will have to start again several times for each fraction.

1. $\frac{1}{2}$ **2.** $\frac{1}{4}$ **3.** $\frac{1}{3}$ **4.** $\frac{1}{5}$

5. $\frac{1}{8}$ **6.** $\frac{1}{9}$ **7.** $\frac{1}{10}$ **8.** $\frac{1}{6}$

Use your cut-out shares to check your answers.

We can make sets of equivalent fractions by dividing, too.

e.g.
$$\frac{8}{16} = \frac{2}{4} = \frac{1}{2}$$
$\div 4 \quad \div 2$
$\div 4 \quad \div 2$

We must always divide both the top and bottom numbers by the same amount. Look for numbers that are factors of both top and bottom numbers.

Making sets of equivalent fractions by dividing is called **cancelling**.

When we cannot divide any more we say the fraction is in its **lowest terms**.

This happens when we can find no more factors common to both the top and the bottom numbers.

Exercise 4

Cancel these fractions into their lowest terms.
The first one has been done for you.

1.
$$\frac{8}{12} = \frac{4}{6} = \frac{2}{3}$$
$\div 2 \quad \div 2$
$\div 2 \quad \div 2$

We cannot divide any more as there are no more common factors.

Now try the rest.

2. $\frac{4}{8}$	**3.** $\frac{6}{12}$	**4.** $\frac{3}{12}$	**5.** $\frac{4}{12}$
6. $\frac{6}{8}$	**7.** $\frac{9}{12}$	**8.** $\frac{5}{10}$	**9.** $\frac{5}{15}$
10. $\frac{2}{4}$	**11.** $\frac{3}{9}$	**12.** $\frac{6}{18}$	**13.** $\frac{4}{16}$
14. $\frac{5}{20}$	**15.** $\frac{8}{24}$	**16.** $\frac{10}{15}$	**17.** $\frac{8}{12}$

Don't blame the calculator

It is quite easy to make mistakes using a calculator.
It is usually not the calculator's fault.
We will make fewer mistakes if we check our calculations.

Checking addition

43 + 16 is the same as 16 + 43.
So we can do addition sums in any order.
To check an addition, do it in the reverse order.

Example

13 + 59 + 77 + 12

First add it up in order: 13 + 59 + 77 + 12 = 161 } same
Then add it up in reverse: 12 + 77 + 59 + 13 = 161 } answer

Exercise 1

Do these additions.
Check them by adding in reverse.

1. 13 + 21	**5.** 28 + 96 + 14	**9.** 34 + 58 + 211
2. 15 + 17 + 42	**6.** 128 + 3 + 15 + 6	**10.** 1921 + 66 + 11
3. 30 + 16 + 124	**7.** 1 + 2 + 3 + 4	**11.** 368 + 1450 + 124
4. 67 + 56 + 3	**8.** 6 + 17 + 196	**12.** 12 + 23 + 34 + 45

Checking multiplication

Multiplication can also be done in any order.
So we can also check multiplications by doing them in reverse.

Example

$54 \times 4 \times 12$

First multiply it in order: $54 \times 4 \times 12 = 2592$ } same
Then multiply it in reverse: $12 \times 4 \times 54 = 2592$ } answer

Exercise 2

Do these multiplications and check them.

1. 3×5 **5.** $2 \times 3 \times 5 \times 7$ **9.** $5 \times 6 \times 3 \times 18$
2. $4 \times 6 \times 2$ **6.** $3 \times 4 \times 5$ **10.** $6 \times 6 \times 4 \times 2$
3. $5 \times 8 \times 7$ **7.** $11 \times 9 \times 6$ **11.** $3 \times 5 \times 3 \times 2$
4. $24 \times 2 \times 3$ **8.** $12 \times 4 \times 2 \times 3$ **12.** $4 \times 2 \times 4 \times 2 \times 8$

Checking subtraction

$36 - 7 = 29$.

So $29 + 7 = 36$.

To check a subtraction, add the answer to the number you took away.
You should get the number you started with.
Don't forget to write the answer down while it is on the display!

Example

$49 - 23$

| C | 4 | 9 | − | 2 | 3 | = | $26.$ |

write down the answer

| + | 2 | 3 | = | $49.$ |

this is what you started with

Exercise 3

Work out these subtractions and check them.

1. $35 - 19$ **5.** $196 - 80$ **9.** $298 - 106$ **13.** $52 - 16$
2. $106 - 30$ **6.** $430 - 197$ **10.** $154 - 97$ **14.** $981 - 56$
3. $84 - 56$ **7.** $56 - 49$ **11.** $27 - 15$ **15.** $1000 - 684$
4. $27 - 9$ **8.** $127 - 114$ **12.** $69 - 22$ **16.** $250 - 194$

Checking division

$35 \div 7 = 5$.

So $5 \times 7 = 35$.

To check a division, multiply the answer by the number you divided by.
You should get the number you started with.

Example

$155 \div 5$

| C | 1 | 5 | 5 | ÷ | 5 | = | | *31.* |

▲
write down the answer

| × | 5 | = | | *155.* |

▲
this is what you started with

Exercise 4

Work out these divisions and check them.

1. $48 \div 3$
2. $63 \div 9$
3. $126 \div 6$
4. $205 \div 41$
5. $340 \div 4$

6. $288 \div 8$
7. $243 \div 9$
8. $156 \div 13$
9. $351 \div 27$
10. $720 \div 20$

11. $260 \div 4$
12. $308 \div 14$
13. $273 \div 3$
14. $138 \div 6$
15. $231 \div 7$
16. $1435 \div 35$

Exercise 5

Here is a mixture of sums.
Work them out and check them.

1. 12×14
2. $135 + 191$
3. $256 - 182$
4. $147 \div 3$
5. $216 \div 9$

6. $156 + 19 + 27$
7. 46×21
8. $384 - 299$
9. $21 + 42 + 84$
10. $318 - 64$

11. $441 \div 21$
12. $3 \times 16 \times 4$
13. $156 \times 2 \times 5$
14. $417 - 36$
15. $375 \div 15$
16. $84 + 1001 + 315$

How long to wait?

School ends at 3.35.
Darren's watch reads 2.25.
How long is it to the end of school?

`02:25`

From 2.25 to 3.25 is one hour.
From 3.25 to 3.35 is ten minutes.
So school will end in one hour and ten minutes.

Exercise 1

1. `10:00` Here is Darren's watch again.
How long is it to

(a) 11.15 (b) 12.45 (c) 1.27

2. `12:45` Look at Darren's watch now.
How long is it to

(a) 7.15 (b) 3.30 (c) 4.25

3. `03:25` How long is it to

(a) 4.30 (b) 5.15 (c) 8.35

(d) 9.50 (e) 6.40 (f) 4.55

4. `09:50` How long is it to

(a) 10.10 (b) 11.40 (c) 10.45

(d) 11.15 (e) 12.35 (f) 2.55

5. `11:10` How long is it to

(a) 12.05 (b) 3.20 (c) 1.45

(d) 5.55 (e) 11.47 (f) 2.40

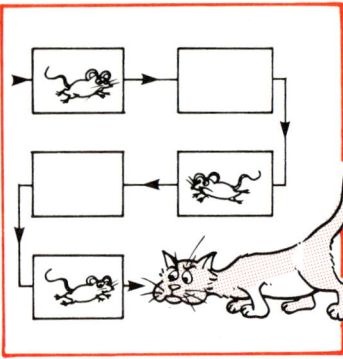

Flow diagrams

Flow diagrams are used to give instructions.
Follow the arrows to find the answer.

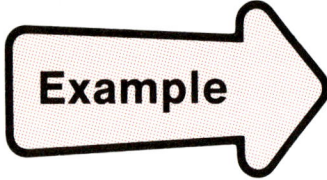

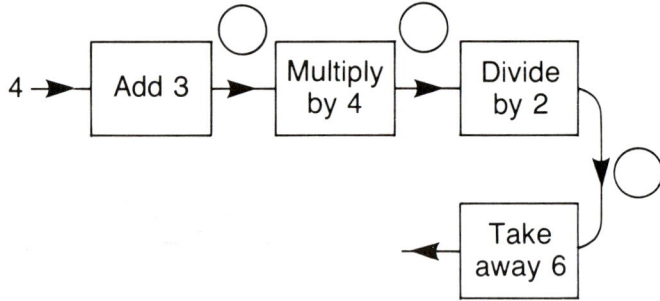

4 → | Add 3 | → ○ → | Multiply by 4 | → ○ → | Divide by 2 | → ○ → | Take away 6 | ←

We fill in the results of each sum in the circles.

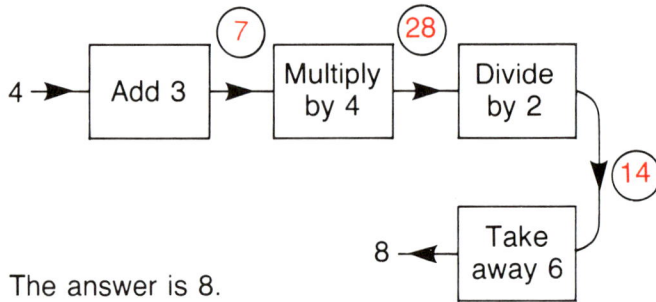

4 → | Add 3 | → (7) → | Multiply by 4 | → (28) → | Divide by 2 | → (14) → | Take away 6 | → 8

The answer is 8.

Exercise 1

Copy these flow diagrams into your book.
Follow the arrows to the final answer.
Fill in the results of each sum in the circles.

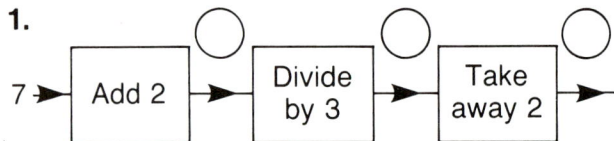

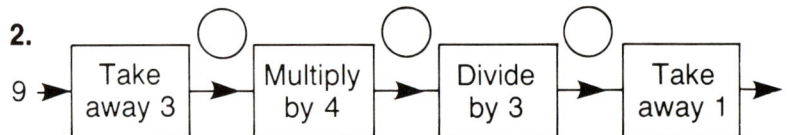

1.

7 → | Add 2 | → ○ → | Divide by 3 | → ○ → | Take away 2 | → ○ →

2.

9 → | Take away 3 | → ○ → | Multiply by 4 | → ○ → | Divide by 3 | → ○ → | Take away 1 | →

3.

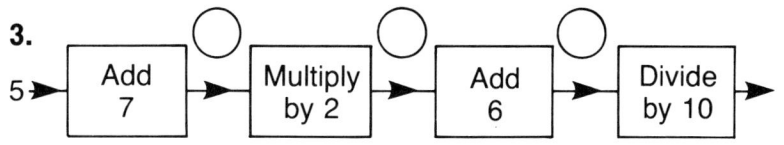

5 → | Add 7 | → ○ → | Multiply by 2 | → ○ → | Add 6 | → ○ → | Divide by 10 | →

4.

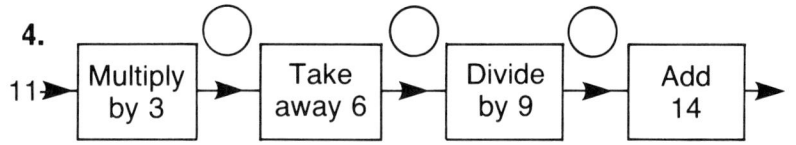

11 → | Multiply by 3 | → ○ → | Take away 6 | → ○ → | Divide by 9 | → ○ → | Add 14 | →

5.

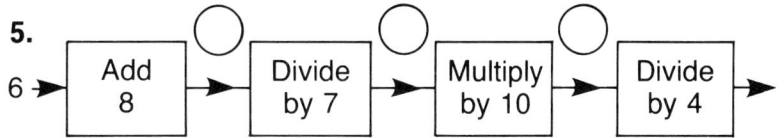

6 → | Add 8 | → ○ → | Divide by 7 | → ○ → | Multiply by 10 | → ○ → | Divide by 4 | →

6.

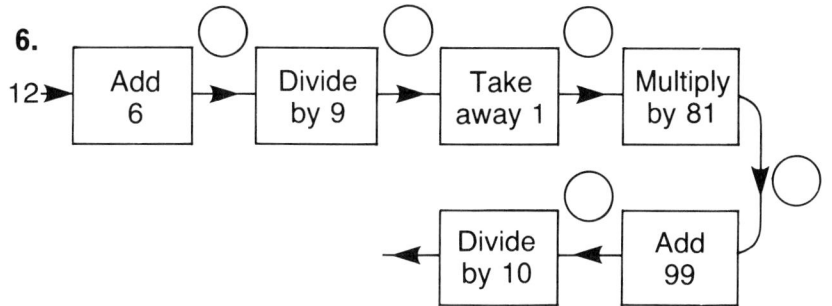

12 → | Add 6 | → ○ → | Divide by 9 | → ○ → | Take away 1 | → ○ → | Multiply by 81 | → ○

← | Divide by 10 | ← | Add 99 | ← ○

7.

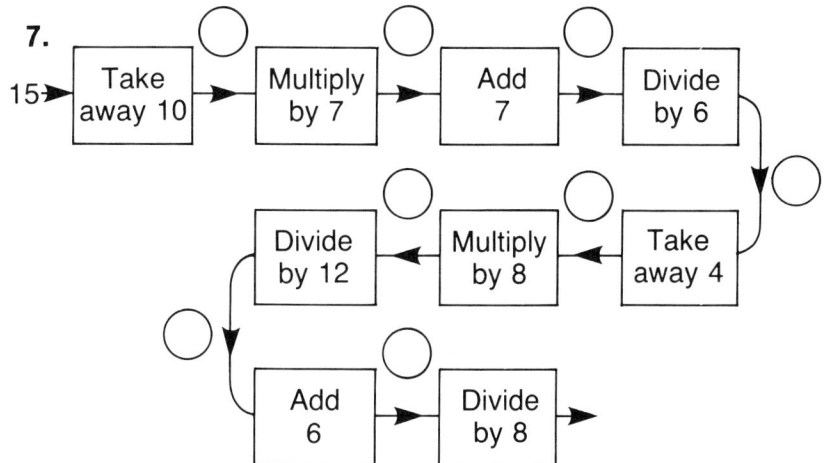

15 → | Take away 10 | → ○ → | Multiply by 7 | → ○ → | Add 7 | → ○ → | Divide by 6 | → ○

← | Divide by 12 | ← ○ ← | Multiply by 8 | ← ○ ← | Take away 4 |

○ → | Add 6 | → ○ → | Divide by 8 | →

We do not always have to start with the same number.
Here is a flow diagram without a starting number.

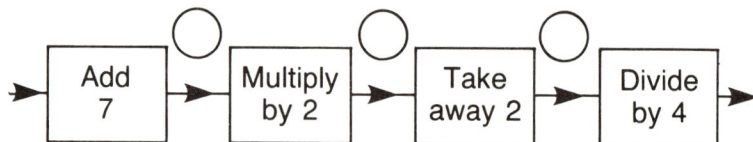

| Add 7 | ○ | Multiply by 2 | ○ | Take away 2 | ○ | Divide by 4 |

Ovid used 4 as a starting number.
He wrote:

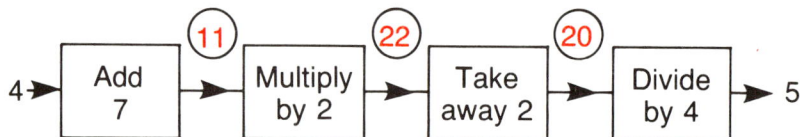

4 → | Add 7 | (11) | Multiply by 2 | (22) | Take away 2 | (20) | Divide by 4 | → 5

Tricia used 12 as her starting number.
She wrote:

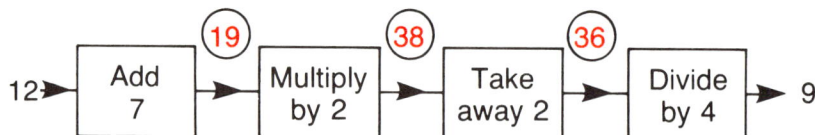

12 → | Add 7 | (19) | Multiply by 2 | (38) | Take away 2 | (36) | Divide by 4 | → 9

Exercise 2

Here are some more flow diagrams.
For each, use (a) 2 (b) 4 (c) 8 as the starting number.

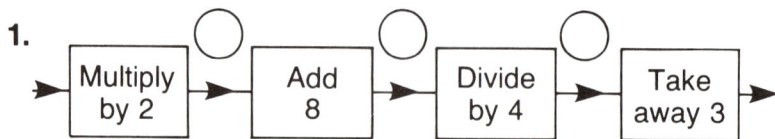

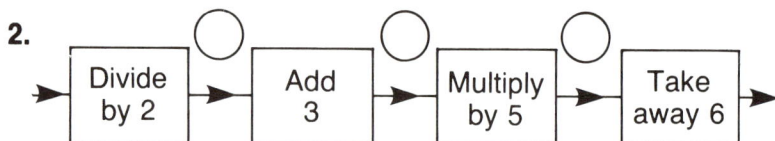

1.

| Multiply by 2 | ○ | Add 8 | ○ | Divide by 4 | ○ | Take away 3 |

2.

| Divide by 2 | ○ | Add 3 | ○ | Multiply by 5 | ○ | Take away 6 |

3.

Multiply by 3 → ◯ → Take away 2 → ◯ → Multiply by 2 → ◯

Divide by 4 → ◯ → Add 7 ←

4.

Multiply by 6 → ◯ → Add 9 → ◯ → Divide by 3

Multiply by 5 → ◯ → Take away 11 ←

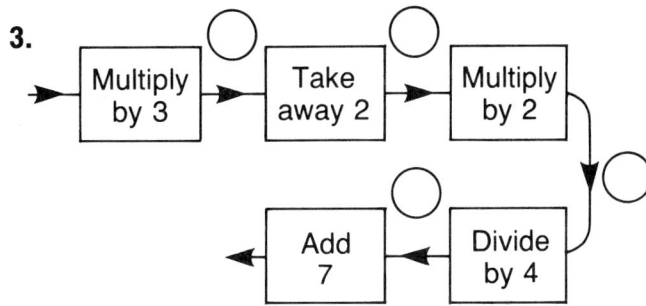

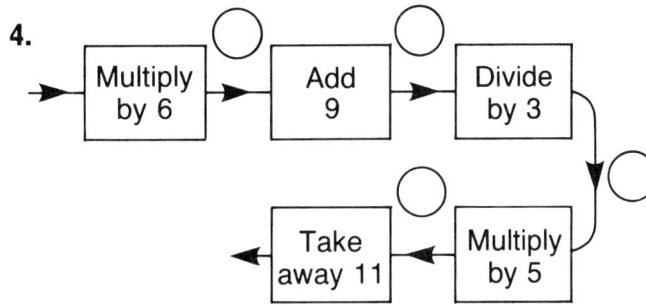

Now use **(a)** 3 **(b)** 5 **(c)** 15 as the starting number in these:

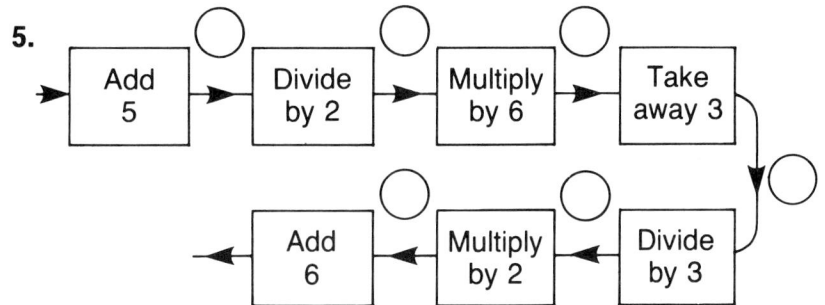

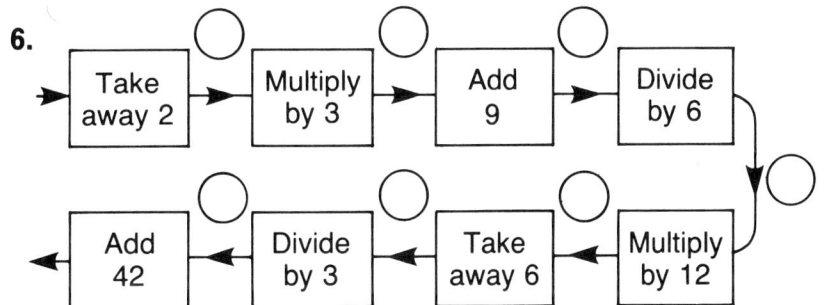

5.

Add 5 → ◯ → Divide by 2 → ◯ → Multiply by 6 → ◯ → Take away 3

Divide by 3 → ◯ → Multiply by 2 → ◯ → Add 6 ←

6.

Take away 2 → ◯ → Multiply by 3 → ◯ → Add 9 → ◯ → Divide by 6

Multiply by 12 → ◯ → Take away 6 → ◯ → Divide by 3 → ◯ → Add 42 ←

63

This is my best angle.

Angles

When we measure **angles** we measure in turns.

The minute hand starts at 12 and turns until it has reached 12 again.
We say it has gone one full turn.

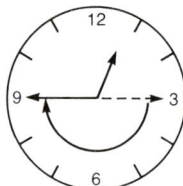

In these two clocks the minute hand has gone one half turn.

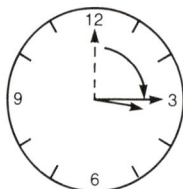

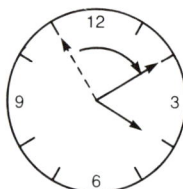

In these two clocks the minute hand has gone one quarter turn.

A turn in the same direction as a clock is called a clockwise turn.
A turn the other way is anticlockwise.

Exercise 1

A. Look at the clocks below.
Say whether the minute hand has gone **(a)** one full turn **(b)** one half turn **(c)** one quarter turn.

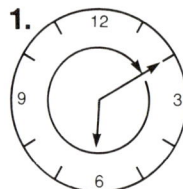

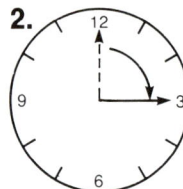

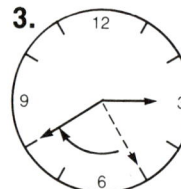

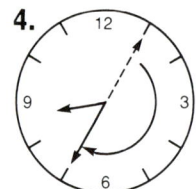

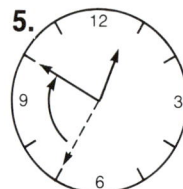

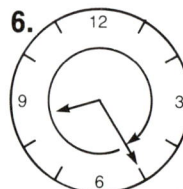

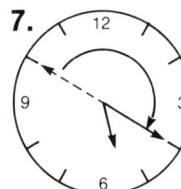

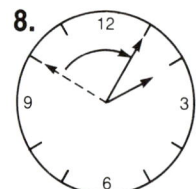

B. Look at the diagrams below.
  For each one say **(a)** how much of a turn it is
  (b) if it is clockwise or anticlockwise

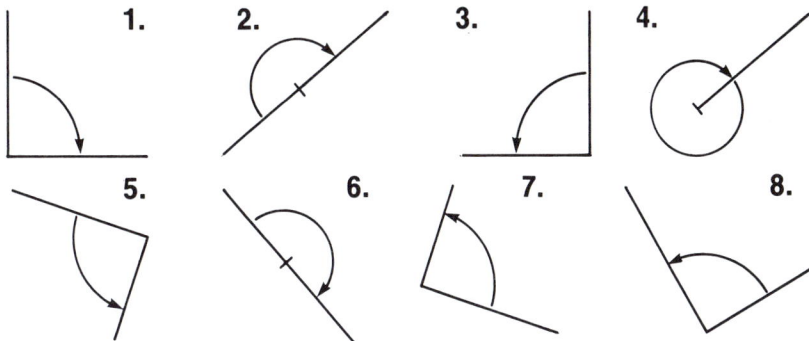

A quarter turn is a special kind of angle.
We call it a **right angle**.

We use a special sign to mark a right angle:

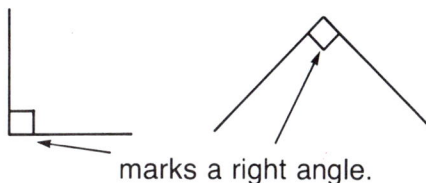

marks a right angle.

Exercise 2

Look at the angles below.
For each one, say whether it is   **(a)** a right angle
(b) more than a right angle   **(c)** less than a right angle.

The first one has been done for you.

1.

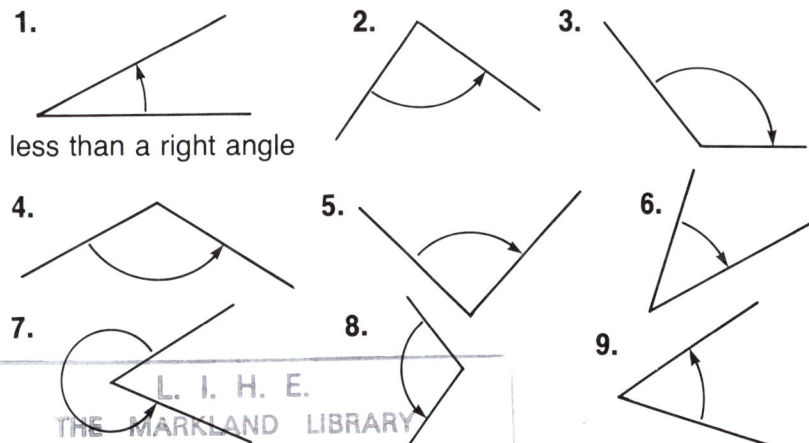

less than a right angle

2.

3.

4.

5.

6.

7.

8.

9.

Dividing

All present and correct

When we divide whole numbers we have to be careful to fill in all the columns.

You can see why if we do a division written in columns.

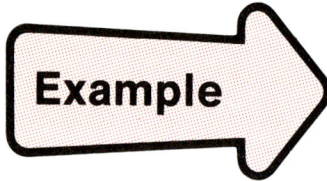

Example

$621 \div 3$

HUNDREDS	TENS	ONES
2	0	7
6	2	1

3

make sure a zero is written in to fill the gap

$621 \div 3 = 207$

Exercise 1

A. Do these division sums.
Take care not to miss out any columns.

1. $306 \div 3$

2. $416 \div 4$

3. $214 \div 2$

4. $108 \div 6$

5. $105 \div 7$

6. $152 \div 8$

7. $315 \div 9$

8. $880 \div 4$

9. $535 \div 5$

10. $270 \div 9$

B. Work out the answers to these general questions.

1. A school buys 75 pocket calculators. Each calculator takes 2 batteries. How many batteries do they need altogether?

2. A shop has milk delivered in crates. Each crate holds 36 bottles. Three crates are delivered. How many bottles are there?

3. The total wage bill for a workshop is £602. There are 7 workers. If they all get paid the same amount, how much does each one get?

4. Jason gets paid £80 for a five-day week. How much is he paid per day?
He works 8 hours a day. How much does he get paid per hour?

Check-up 3

1. Find four factors of each of these numbers:

(a) 28   **(b)** 18   **(c)** 54

2. Write these in shorthand:

(a) $5 \times p$ **(b)** $3 \times a \times b$

3. Work these out:

(a) $4p + 5p$ **(b)** $7b - 2b$

4. Use a calculator to change these fractions to decimals:

(a) $\dfrac{3}{4}$ **(b)** $\dfrac{4}{5}$ **(c)** $2\dfrac{7}{10}$

5. Draw **(a)** a rectangle   **(b)** a parallelogram   **(c)** a kite.

6. Cancel these fractions to their lowest terms:

(a) $\dfrac{6}{8}$ **(b)** $\dfrac{5}{10}$ **(c)** $\dfrac{2}{4}$

7. Use 7 as the starting number in this flow diagram:

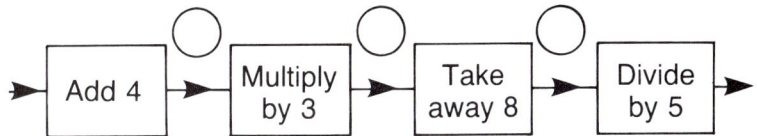

8. For each angle below, say whether it is more or less than a right angle:

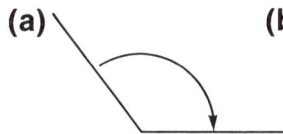

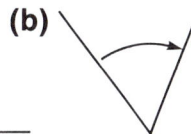

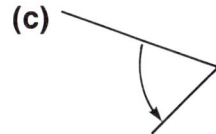

(a)   **(b)**   **(c)**

9.

How long is it to:

(a) 10.00   **(b)** 10.40   **(c)** 12.15

10. Work these out:

(a) $306 \div 3$ **(b)** 864×8 **(c)** $2418 \div 6$

You can have $\frac{1}{4}$
You can have $\frac{2}{8}$
and you can have $\frac{4}{16}$

Adding and subtracting fractions

Alice, Dermot and Maxine have a quarter of a cake each. How much do they have altogether?

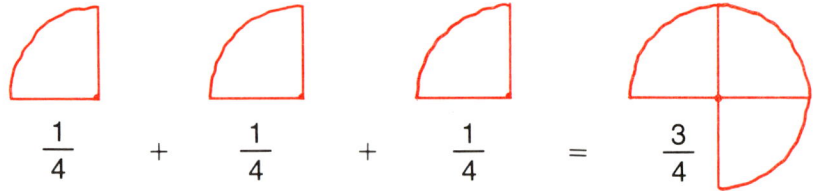

$$\frac{1}{4} \quad + \quad \frac{1}{4} \quad + \quad \frac{1}{4} \quad = \quad \frac{3}{4}$$

They have three quarters altogether.

Exercise 1

Work these out the same way. Write the answer in words and figures.

1. $\frac{1}{8} + \frac{1}{8} + \frac{1}{8}$ **3.** $\frac{1}{5} + \frac{3}{5}$ **5.** $\frac{5}{10} + \frac{3}{10}$

2. $\frac{1}{3} + \frac{2}{3}$ **4.** $\frac{3}{8} + \frac{1}{8}$ **6.** $\frac{1}{3} + \frac{1}{3} + \frac{2}{3}$

When we add fractions, we have to make sure we have fractions that are all the same type.
We use equivalent fractions.

$$\frac{1}{4} + \frac{1}{8}$$

Example

We can make the quarter into eighths using equivalent fractions.

$$\frac{1}{4} \overset{\times 2}{\underset{\times 2}{=}} \frac{2}{8}$$

So the sum is now $\frac{2}{8} + \frac{1}{8} = \frac{3}{8}$

68

Exercise 2

Use equivalent fractions to work these out.
Change the larger share so it is the same sort as the smaller one.
The first one has been done for you.

1. $\dfrac{1}{6} + \dfrac{2}{3}$

Change the thirds into sixths:

$$\overset{\times 2}{\dfrac{2}{3}} = \dfrac{4}{6} \quad {\scriptstyle \times 2}$$

We have $\dfrac{1}{6} + \dfrac{4}{6} = \dfrac{5}{6}$

Now try these. Check that the answers are in their lowest terms.

2. $\dfrac{1}{2} + \dfrac{1}{4}$ **5.** $\dfrac{3}{8} + \dfrac{1}{4}$ **8.** $\dfrac{5}{8} + \dfrac{3}{4}$

3. $\dfrac{2}{5} + \dfrac{1}{10}$ **6.** $\dfrac{3}{5} + \dfrac{3}{10}$ **9.** $\dfrac{7}{10} + \dfrac{2}{5}$

4. $\dfrac{5}{6} + \dfrac{1}{3}$ **7.** $\dfrac{5}{8} + \dfrac{1}{16}$ **10.** $\dfrac{2}{3} + \dfrac{4}{9}$

Sometimes we cannot change the larger share so it is the same sort as the smaller one.
We have to change both fractions.

Example

$\dfrac{2}{3} + \dfrac{3}{4}$

Thirds and quarters can both be changed into twelfths.

$$\overset{\times 4}{\dfrac{2}{3}} = \dfrac{8}{12} \qquad \overset{\times 3}{\dfrac{3}{4}} = \dfrac{9}{12}$$

We have $\dfrac{8}{12} + \dfrac{9}{12} = \dfrac{17}{12}$

$\dfrac{17}{12}$ is more than one whole one.

We change it to whole numbers and fractions.

$\dfrac{17}{12} = 1\dfrac{5}{12}$

Exercise 3

Work these out. If necessary, cancel your answers and change them to whole numbers and fractions.

1. $\dfrac{1}{4} + \dfrac{2}{3}$ **3.** $\dfrac{2}{5} + \dfrac{1}{4}$ **5.** $\dfrac{4}{5} + \dfrac{1}{2}$

2. $\dfrac{1}{2} + \dfrac{1}{3}$ **4.** $\dfrac{2}{3} + \dfrac{1}{4}$ **6.** $\dfrac{3}{5} + \dfrac{2}{3}$

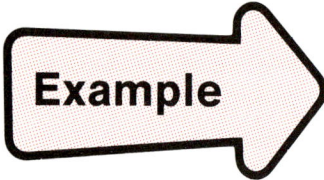

Subtracting fractions is done the same way as addition. Instead of adding the shares, we subtract.

Example

$$\dfrac{4}{5} - \dfrac{1}{3}$$

Fifths and thirds can be changed into fifteenths.

$$\dfrac{4}{5} \overset{\times 3}{\underset{\times 3}{=}} \dfrac{12}{15} \qquad \dfrac{1}{3} \overset{\times 5}{\underset{\times 5}{=}} \dfrac{5}{15}$$

We have $\dfrac{12}{15} - \dfrac{5}{15} = \dfrac{7}{15}$

Exercise 4

Now try these:

1. $\dfrac{1}{2} - \dfrac{1}{4}$ **4.** $\dfrac{1}{2} - \dfrac{3}{8}$ **7.** $\dfrac{4}{5} - \dfrac{7}{10}$

2. $\dfrac{3}{4} - \dfrac{3}{8}$ **5.** $\dfrac{2}{3} - \dfrac{3}{5}$ **8.** $\dfrac{3}{8} - \dfrac{1}{4}$

3. $\dfrac{5}{6} - \dfrac{1}{2}$ **6.** $\dfrac{3}{4} - \dfrac{5}{12}$ **9.** $\dfrac{5}{6} - \dfrac{2}{3}$

Parallelograms

Parallelograms are shapes made of two sets of parallel lines.

Here are some parallelograms:

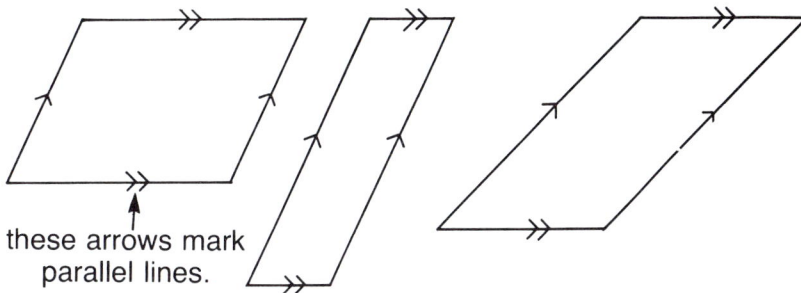

these arrows mark parallel lines.

Opposite sides of parallelograms are parallel.

Here are some special kinds of parallelogram. Copy them into your book.

A **rectangle** is a special kind of parallelogram.

Its opposite sides are parallel.

All the sides of this parallelogram are the same length. It is called a **rhombus**.

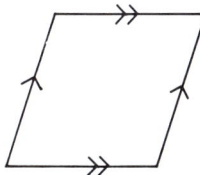

A **square** is a special kind of parallelogram.

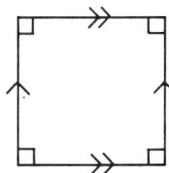

It is also a special kind of rhombus.

Decimals

Do the sum $1 \div 10$ on the calculator.

$1 \div 10 = 0.1$

$\dfrac{1}{10} = 0.1$ one share out of 10 $= 0.1$

one tenth $= 0.1$

We call the first column after the decimal point the TENTHS column.

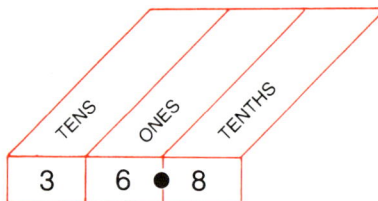

e.g. 36.8

TENS	ONES	TENTHS
3	6 ●	8

Earlier in the book we called this column the 10ps column.

This is because   £1 $\div$ 10 = 10p

There are ten 10ps in £1

10p is $\dfrac{1}{10}$ of £1

Do the sum $1 \div 100$ on the calculator.

$1 \div 100 = 0.01$

$\dfrac{1}{100} = 0.01$ one share out of 100 $= 0.01$

one hundredth $= 0.01$

We call the second column after the decimal point the HUNDREDTHS column.

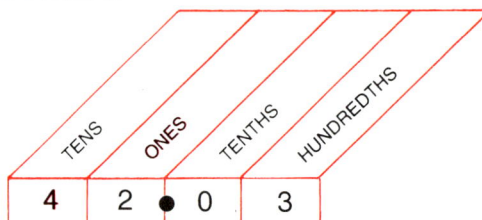

e.g. 42.03

TENS	ONES	TENTHS	HUNDREDTHS
4	2 ●	0	3

This is the column we called the 1ps column.

£1 $\div$ 100 = 1p

There are 100 pence in £1

1p is $\dfrac{1}{100}$ of £1

Decimal numbers are just like pounds and pence.
We use them because they are often simpler than fractions.
All sums on the calculator must be done using decimals.

When we multiply or divide decimal numbers by 10, 100 and 1000, they move columns in the same way as whole numbers.

The decimal point will then be between different numbers.

$3.6 \div 10$

3.6

$3.6 \div 10$

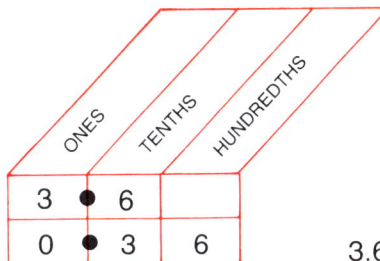

$3.6 \div 10 = 0.36$

0.06×100

0.06

0.06×100

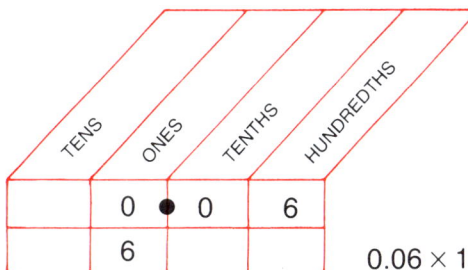

$0.06 \times 100 = 6$

You don't need to write in the decimal point if there is no number after it.

Exercise 1

Work out these multiplications and divisions in the same way.

1. 3.6×10	**6.** 2.74×10	**11.** 34.5×10
2. $0.2 \div 10$	**7.** 9.33×10	**12.** 2.21×10
3. $34.2 \div 10$	**8.** $46.3 \div 10$	**13.** $22.1 \div 10$
4. $190 \div 100$	**9.** $27 \div 100$	**14.** $221 \div 100$
5. 4.63×100	**10.** $1960 \div 1000$	**15.** $100 \div 1000$

Triangles

Triangles are shapes with three sides and three angles.

Different triangles have different names.

The names describe the triangles. They tell us about the sides or angles of the triangle.

The names of these triangles tell us about the sides:

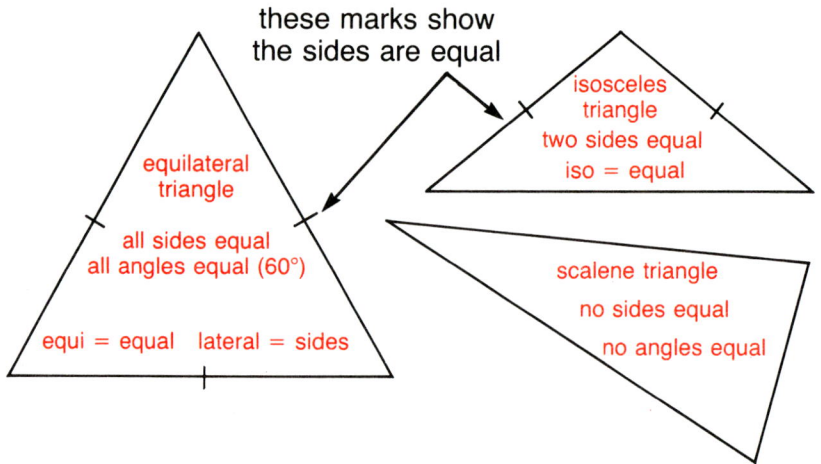

these marks show
the sides are equal

equilateral
triangle

all sides equal
all angles equal (60°)

equi = equal   lateral = sides

isosceles
triangle
two sides equal
iso = equal

scalene triangle

no sides equal

no angles equal

The names of these triangles tell us about the angles:

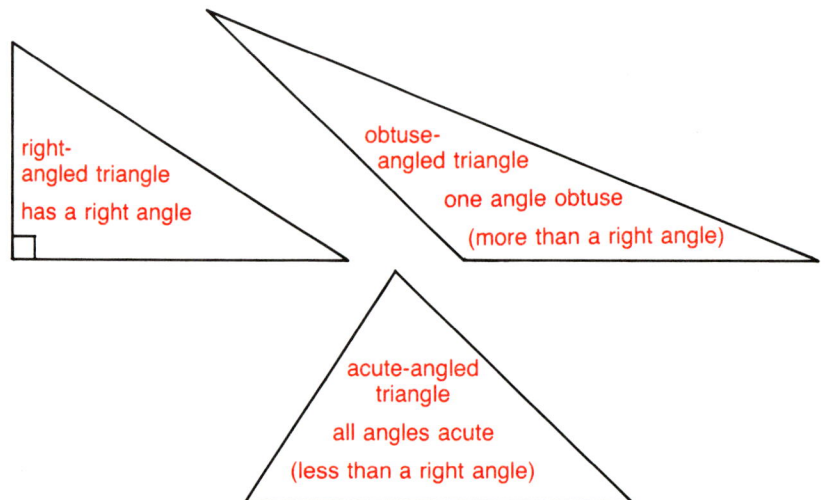

right-
angled triangle

has a right angle

obtuse-
angled triangle

one angle obtuse

(more than a right angle)

acute-angled
triangle

all angles acute

(less than a right angle)

Exercise 1

Name the triangles below.
Use the information on page 74 to help you.
Some can be named in more than one way.

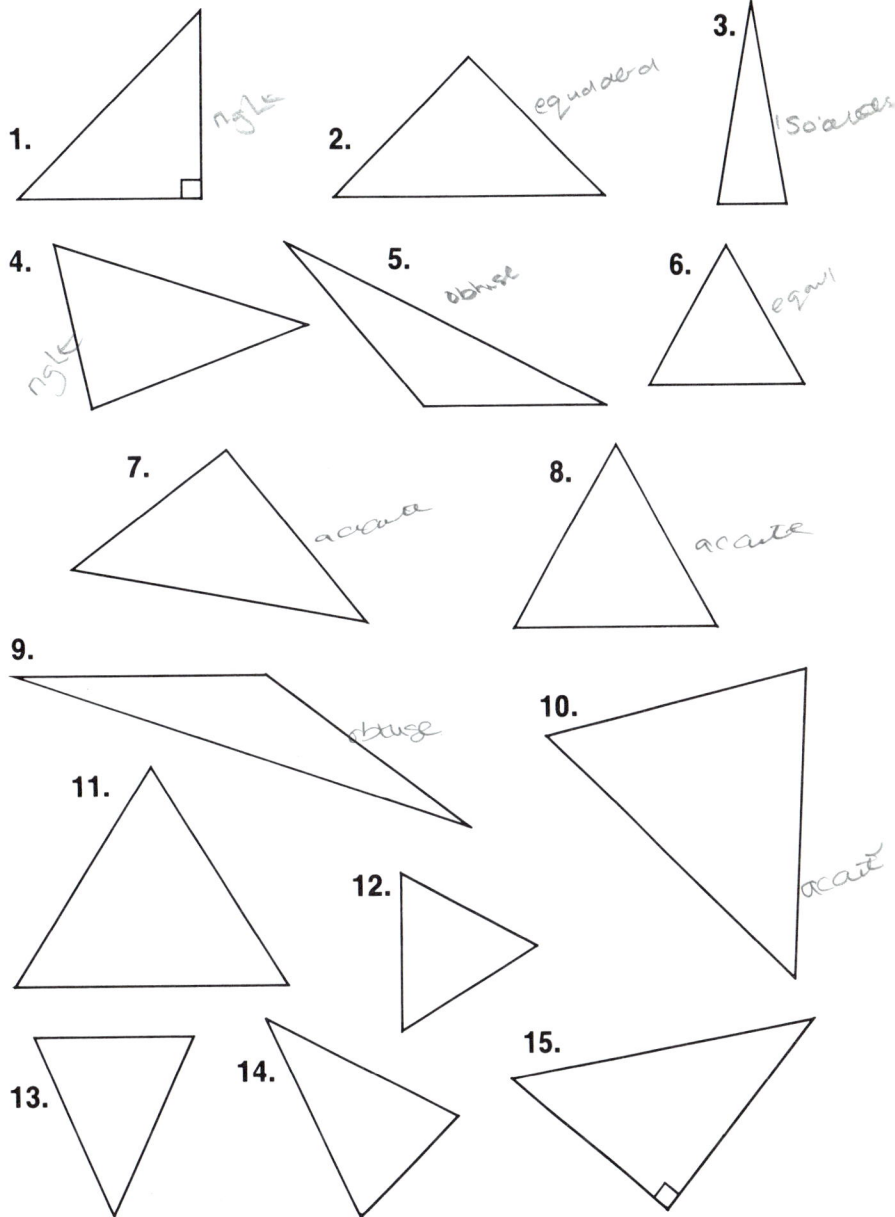

1. rngle

2. equalderol

3. Isoceles

4. ngle

5. obhuse

6. egaul

7. acaua

8. acaute

9. rtuse

10. acait

11.

12.

13.

14.

15.

Surveys

Curtis and his friends do a traffic survey.
They record how many vehicles go past the school gates in twenty minutes.
Each person records a different sort of vehicle.
They use tallies to record their results.

Curtis records cars.
Each time a car goes by he makes a mark like this: I
When he gets to five he crosses through the other four marks like this: ∥∥∥

Here are Curtis's results:

vehicle	tally	number
cars	∥∥∥ ∥∥∥ ∥∥∥ ∥∥∥ ∥∥∥ ∥	27

Exercise 1

1. Curtis and his friends make a table of their results.
Copy the table into your book and complete it.

vehicle	tally	number
cars	∥∥∥ ∥∥∥ ∥∥∥ ∥∥∥ ∥∥∥ ∥	27
lorries	∥∥	
buses	∥∥∥ ∥∥∥	
taxis	I	
vans		6
motor cycles		8
bicycles		13
		total:

2. Maxine and her friends do a survey of how people in their class come to school.
Copy and complete their table.

method	tally	number
car	∥∥	
bus		7
train	I	
bicycle	∥∥∥ ∥∥	
walk		11
		total:

76

Tallies are useful for other things too.

Here are the scores of people in Curtis's class in a test.
There are 28 pupils in the class.

```
 3   6   4   2   3    3   4
 1   4   8   9   8    9   3
10   9   6   6   2    1   5
 5   4   2   3   9   10   6
```

The teacher wants to know how many people got each score.
She tallies off the marks one by one on a tally chart.
Here is the chart after she has done the first row of scores:

score	tally	number
1		
2	I	
3	III	
4	II	
5		
6	I	
7		
8		
9		
10		
		total:

Exercise 2

1. Copy and complete the teacher's tally chart.
You can check the total because you know there are 28 pupils in the class.

2. Maxine did a survey of the size of families of 32 people.
Here are her results.

```
3   4   2   6   2   4   6   4
4   2   3   3   5   4   5   5
6   7   5   5   4   2   4   3
2   7   9   3   4   5   3   3
```

Make a tally chart of Maxine's results.
What was the most common size of family?
What was the least common size of family?

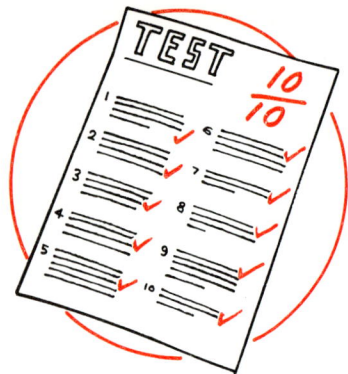

Curtis's teacher draws a **bar chart** to display her results.

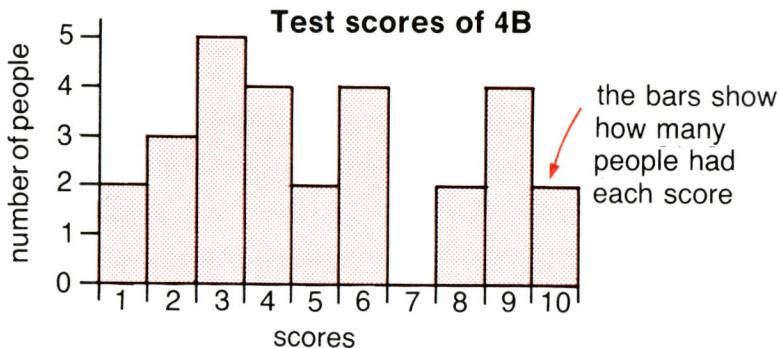

Test scores of 4B

the bars show how many people had each score

We use bar charts to compare numbers of things.

When you draw a bar chart, remember:

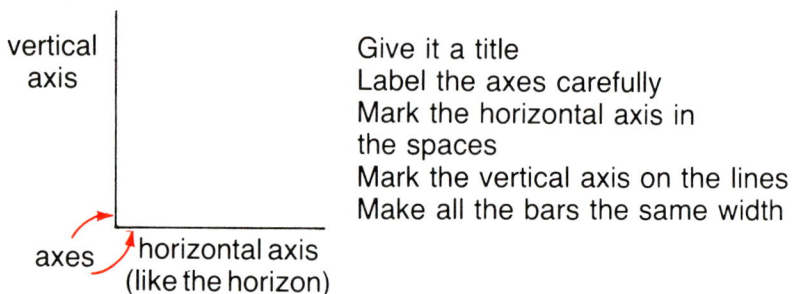

vertical axis

axes — horizontal axis (like the horizon)

Give it a title
Label the axes carefully
Mark the horizontal axis in the spaces
Mark the vertical axis on the lines
Make all the bars the same width

Exercise 3

1. Draw a bar chart to display the results of Maxine's survey.

2. Alice did a survey of shoe sizes in her class.
She displayed her results on a bar chart.

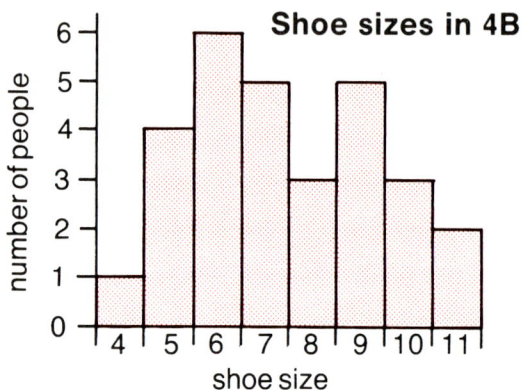

Shoe sizes in 4B

(a) What was the most common shoe size?
(b) What was the least common shoe size?
(c) How many people took size 8?
(d) How many people took size 5?

When we draw bar charts of large numbers of things, we need to use a different scale for the vertical axis.

Here are the results of a survey of colours of cars in a car park:

colour	number
red	35
blue	23
yellow	15
green	27
white	30
black	4
	total: 134

We mark off every 5 along the vertical axis:

Colours of cars in a car park

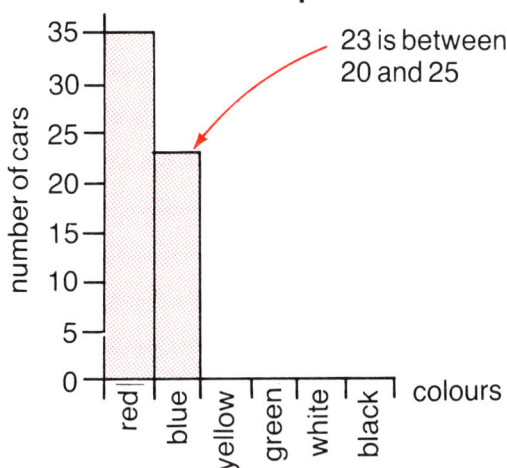

23 is between 20 and 25

Exercise 4

1. Copy and complete the bar chart of car colours.

2. Here are the results of a survey of the favourite sports of 150 people:

sport	number
tennis	15
football	42
hockey	9
cricket	18
netball	25
basketball	24
swimming	17
	total: 150

Draw a bar chart to show these results. Mark off every 5 on the vertical axis.

We can also use bar charts to show how much of something there is at different times and places.
We use different scales for these charts too.

3. Here is a bar chart showing rainfall over a year:

Rainfall over one year

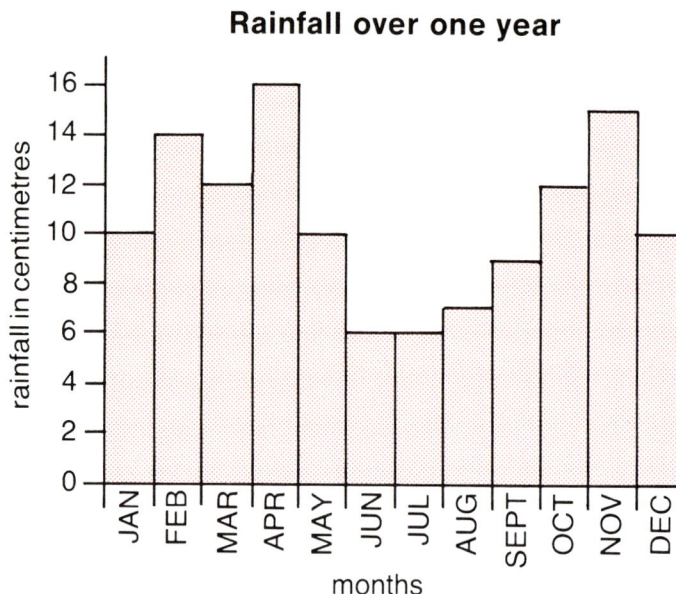

(a) Which month did it rain most?
(b) Which months did it rain least?
(c) How much rain fell in March?
(d) How much rain fell in August?
(e) How much more rain fell in October than in June?
(f) How much less rain fell in December than in February?

4. Here is a table showing the number of people visiting a squash club each day.

day	numbers
Monday	30
Tuesday	70
Wednesday	25
Thursday	40
Friday	65
Saturday	45
Sunday	80

Draw a bar chart to show this data.
Mark off every 10 on the vertical axis.

5. Here is a bar chart showing sales of the Breakaways' hit record 'Smashit' over six weeks.

Sales of 'Smashhit' over six weeks

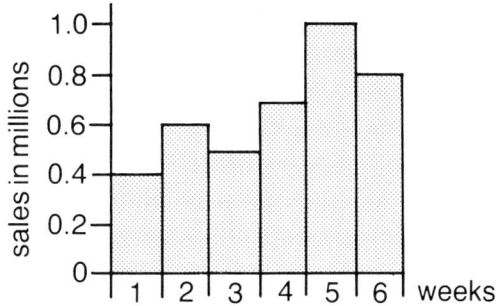

(a) How many records were sold in week 1?
(b) How many records were sold in week 3?
(c) How many more records were sold in week four than in week one?
(d) How much did sales decrease between week five and week six?
(e) How many records were sold altogether during the six weeks?

6. Here is a table showing the amounts raised for different charities by 4G in one term.

charity	amount raised
Oxfam	£7.50
RSPCA	£1.50
Lifeboats	£3.00
Mentally handicapped	£2.75
Cancer research	£5.25
OAPs' Party	£6.50

(a) Draw a bar chart to show this data.
Mark off every 50p on the vertical axis.

(b) How much money was raised altogether?

Which units?

Earlier in this book we used metres (m), centimetres (cm) and millimetres (mm) for measuring lengths.

We used millimetres for very small lengths (the length of a fly)
centimetres for small lengths (the length of a pencil)
metres for medium lengths (the length of a room)

For very large lengths we use kilometres (km).

Exercise 1

Which units would you use to measure these?

1. a pen
2. the distance from Nottingham to London
3. the thickness of a book
4. the width of a room
5. the height of a building
6. the length of a motorway
7. the length of a swimming pool
8. the width of a matchbox
9. the width of your fingernail
10. the height of a desk

The **metre** is the basic unit of length in the metric system. All the other units are based on it.

Cent means hundred.
There are 100 centimetres in 1 metre.

Milli means thousand.
There are 1000 **milli**metres in 1 metre.

So 1 cm = 10 mm

Kilo means thousand, too.
There are 1000 metres in 1 **kilo**metre.

conversion table		
1 km	=	1000 m
1 m	=	100 cm
	=	1000 mm
1 cm	=	10 mm

We can use columns to convert kilometres to metres and back again.

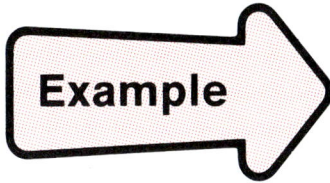

We don't usually use these

km	hm	dam	m	
1 ·	0	0	0	1 km = 1000 m
3 ·	5	0	0	3.5 km = 3500 m
4 ·	5	1	6	4516 m = 4.516 km
0 ·	1	2	7	127 m = 0.127 km

we write in a zero ➡

Example

Exercise 2

Copy these and fill in the gaps:

1. 2000 m = _____ km

2. 1500 m = _____ km

3. 50 000 m = _____ km

4. 3512 m = _____ km

5. 4036 m = _____ km

6. 4 km = _____ m

7. 5.8 km = _____ m

8. 6.421 km = _____ m

9. 10.324 km = _____ m

10. 2.23 km = _____ m

Now do the same with these.
You may need to have another look at the work on metres, centimetres and millimetres.
You will find it on pages 28–29 and page 43.

11. 1.8 m = _____ mm

12. 25 mm = _____ cm

13. 350 cm = _____ m

14. 425 cm = _____ m

15. 14.6 m = _____ cm

16. 3.5 cm = _____ mm

17. 120 m = _____ km

18. 3760 m = _____ km

19. 1.6 km = _____ m

20. 0.62 m = _____ cm

21. 62 cm = _____ mm

22. 18.4 m = _____ cm

23. 18.4 km = _____ m

24. 3500 m = _____ km

25. 650 mm = _____ cm

26. 8.5 cm = _____ mm

Splitting it up

We can multiply large numbers more easily if we split one number into parts.

Example

218 × 45

45 = 40 + 5

So 218 × 45 = 218 × 40 + 218 × 5

218 × 40 = 8720	218	218
218 × 5 = 1090	×5	×4
218 × 45 = 9180	1090	872

218 × 45 = 9810   872 × 10 = 8720

Exercise 1

Work out these multiplications by splitting one number into parts. The first one has been done for you.

1. 255 × 125

125 = 100 + 20 + 5
So 255 × 125 = 255 × 100 + 255 × 20 + 255 × 5

255 × 100 = 25 500	255	255
255 × 20 = 5100	×5	×2
255 × 5 = 1275	1275	510

255 × 125 = 31 875   510 × 10 = 5100

255 × 125 = 31 875

Now you do the rest the same way:

2. 135 × 25   **4.** 375 × 25   **6.** 95 × 575   **8.** 360 × 218
3. 450 × 55   **5.** 125 × 155   **7.** 280 × 66   **9.** 775 × 75

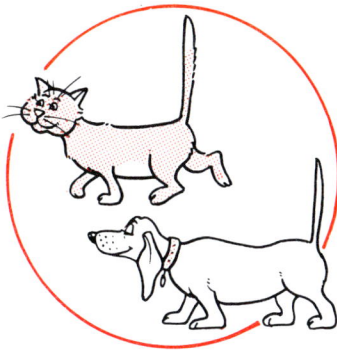

Perpendicular lines

Perpendicular lines are at right angles to each other.

The sides of a table are perpendicular.
So are the corners of this book.

Exercise 1

Copy these drawings into your exercise book.
Colour lines perpendicular to each other the same colour.

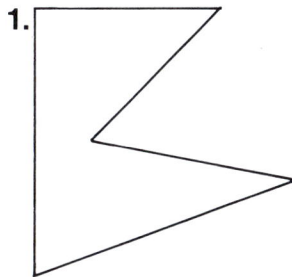

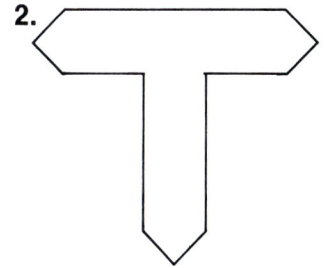

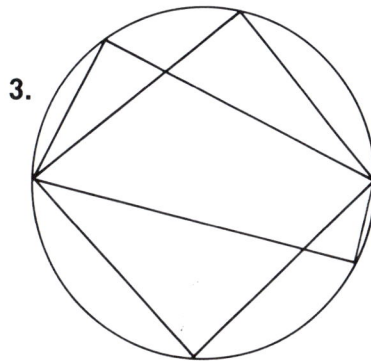

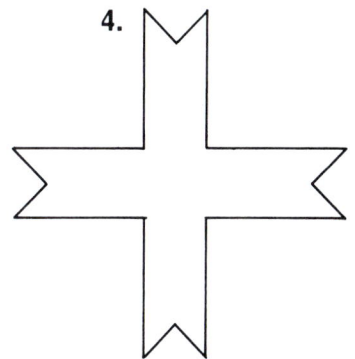

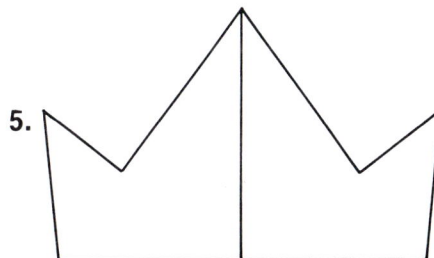

1.

2.

3.

4.

5.

Times means of

3×4 means 3 lots of 4.

In the same way, $\dfrac{1}{2} \times \dfrac{1}{4}$ means $\dfrac{1}{2}$ of $\dfrac{1}{4}$.

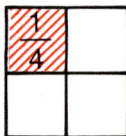 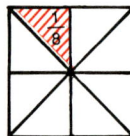

$$\dfrac{1}{4} \qquad\qquad \dfrac{1}{2} \text{ of } \dfrac{1}{4} = \dfrac{1}{8}$$

Exercise 1

Work these out by drawing whole ones and sharing them. The first one has been done for you.

1. $\dfrac{1}{4} \times \dfrac{1}{3} = \dfrac{1}{4}$ of $\dfrac{1}{3}$

$= \dfrac{1}{12}$

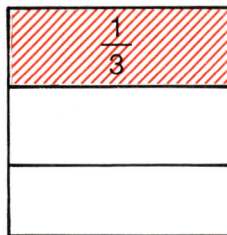 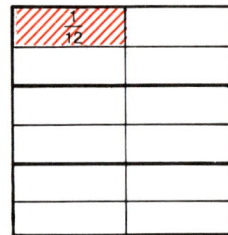

$$\dfrac{1}{3} \qquad\qquad \dfrac{1}{4} \text{ of } \dfrac{1}{3} = \dfrac{1}{12}$$

2. $\dfrac{1}{2} \times \dfrac{1}{2}$

3. $\dfrac{1}{4} \times \dfrac{1}{2}$

4. $\dfrac{1}{3} \times \dfrac{1}{2}$

5. $\dfrac{1}{2} \times \dfrac{1}{3}$

6. $\dfrac{1}{4} \times \dfrac{1}{4}$

7. $\dfrac{1}{3} \times \dfrac{1}{3}$

8. $\dfrac{1}{2} \times \dfrac{1}{5}$

9. $\dfrac{1}{2} \times \dfrac{2}{3}$

10. $\dfrac{1}{4} \times \dfrac{2}{3}$

11. $\dfrac{1}{6} \times \dfrac{1}{2}$

Darren notices something.

'We can work out the answers a quicker way,' he says. 'You get the same answer if you just multiply the numbers together.'

$$\frac{1}{2} \times \frac{1}{4} = \frac{1 \times 1}{2 \times 4} = \frac{1}{8}$$

Multiply top numbers by top numbers and bottom numbers by bottom numbers.

Sometimes the answer is not in its lowest terms.

e.g. $\dfrac{2}{3} \times \dfrac{3}{4} = \dfrac{6}{12}$

We use equivalent fractions to cancel the answer:

$$\frac{6}{12} = \frac{1}{2} \quad \overset{\div 6}{\underset{\div 6}{}}$$

Exercise 2

Work these out using Darren's quick method. Cancel the answers to their lowest terms if you need to.

1. $\dfrac{1}{4} \times \dfrac{1}{3}$

2. $\dfrac{1}{5} \times \dfrac{2}{3}$

3. $\dfrac{3}{4} \times \dfrac{1}{2}$

4. $\dfrac{1}{8} \times \dfrac{1}{4}$

5. $\dfrac{1}{5} \times \dfrac{1}{2}$

6. $\dfrac{2}{3} \times \dfrac{1}{6}$

7. $\dfrac{2}{3} \times \dfrac{5}{6}$

8. $\dfrac{3}{5} \times \dfrac{1}{2}$

9. $\dfrac{3}{4} \times \dfrac{1}{10}$

10. $\dfrac{1}{12} \times \dfrac{2}{3}$

Variable inputs

Maxine does computer studies.
She writes a program to add two numbers.
Here it is:

```
10   INPUT A
20   INPUT B
30   LET C=A+B
40   PRINT A;"+";B;"=";C
50   END
```

Maxine runs her program.
She writes the input   4,5   input = what you type into
                        ↑ ↑   the computer
                        A B

She gets the printout: 4+5=9   printout = what the computer
                          ↑   prints out
                          C

Maxine can run her program with any inputs she likes.

Example

1 input: 3,7
   printout: 3+7=10
2 input: 17,24
   printout: 17+24=41

Exercise 1

A. Run Maxine's program with these inputs.
Write out the printout.

1. 4,8	**4.** 8,63	**7.** 96,18
2. 3,6	**5.** 19,107	**8.** 45,53
3. 17,4	**6.** 1000,101	**9.** 136,27

B. For each input below, write down A, B and C.

1. 3,5	**4.** 47,16	**7.** 11,11
2. 4,9	**5.** 27,148	**8.** 39,41
3. 120,35	**6.** 21,22	**9.** 27,127

Benjamin writes a program to subtract numbers.
Here it is.

```
10   INPUT A
20   INPUT B
30   LET C=A−B
40   PRINT A;"+";B;"=";C
50   END
```

Benjamin runs his program.

input: 8,6
printout: 8−6=2

Exercise 2

A. Run Benjamin's program with these inputs.
Write out the printout.

1. 9,4	**4.** 36,18	**7.** 883,57
2. 16,7	**5.** 26,4	**7.** 360,210
3. 114,27	**6.** 450,97	**9.** 115,115

B. For each input below, write down A, B and C.

1. 18,6	**4.** 63,45	**7.** 590,286
2. 44,12	**5.** 71,18	**7.** 312,147
3. 39,15	**6.** 46,27	**9.** 205,32

> Letters that are used to stand for numbers
> are called **variables**.
>
> We can vary the number they stand for.

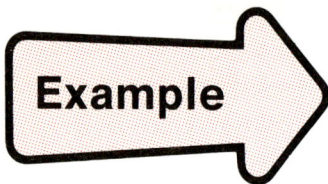

Variables are used in algebra, too.

2a + b

a and b are variables.
We can use any numbers we like.

Remember:
2a = 2 × a

Maxine used a = 4 and b = 5
So 2a + b = 2 × 4 + 5
= 8 + 5
= 13

Benjamin used a = 1 and b = 3
So 2a + b = 2 × 1 + 3
= 2 + 3
= 5

Exercise 3

A. Find the value of 2a + b using these variables:

1. a = 2, b = 3   **4.** a = 3, b = 100
2. a = 6, b = 2   **5.** a = 14, b = 6
3. a = 7, b = 4   **6.** a = 12, b = 35

B. Find the value of 3p + 2q using these variables:

1. p = 3, q = 4   **4.** p = 20, q = 10
2. p = 2, q = 1   **5.** p = 4, q = 30
3. p = 6, q = 3   **6.** p = 11, q = 15

C. Find the value of ab − c using these variables:

1. a = 2, b = 3, c = 4   **4.** a = 16, b = 1, c = 11
2. a = 1, b = 5, c = 3   **5.** a = 9, b = 8, c = 7
3. a = 10, b = 2, c = 9   **6.** a = 4, b = 9, c = 25

Check-up 4

1. Work these out:

   (a) $\dfrac{1}{8} + \dfrac{3}{8}$ **(b)** $\dfrac{1}{6} + \dfrac{1}{3}$ **(c)** $\dfrac{3}{5} - \dfrac{3}{10}$

2. Name these triangles:

  (a)   **(b)**   **(c)**

 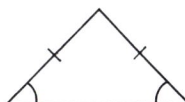 

3. Work these out:

   (a) 4.8×10 **(b)** $20.6 \div 10$ **(c)** 4.3×100

4. Work these out:

   (a) 35×14 **(b)** 150×25

5. Which of these numbers is prime?

   (a) 85   **(b)** 21   **(c)** 29

6. Here are the results of a survey of what magazines and comics people read.

Draw a bar chart to show this data. Mark off every 5 on the vertical axis.

magazine	number
Jilly	57
Teen-Dream	15
Lads Own	20
Falcon	12
Sugar	48
total:	152

7. Draw two special parallelograms.

8. Copy these and fill in the gaps:

   (a) $2500 \, m =$ —————— km   **(b)** $3.65 \, km =$ —————— m

9. Work these out:

   (a) $\dfrac{2}{5} \times \dfrac{1}{4}$ **(b)** $\dfrac{1}{2} \times \dfrac{1}{2}$ **(c)** $\dfrac{1}{3} \times \dfrac{5}{6}$

10. Find the value of $3a - 2b$ using these variables:

   (a) $a = 3, b = 2$ **(b)** $a = 4, b = 5$

Multiplying and dividing decimals

Multiplying and dividing decimals is just like multiplying and dividing money.

Example

4.35 × 6

Write the numbers in columns:

$$\begin{array}{r} 4.35 \\ \times \quad 6 \\ \hline 26.10 \end{array}$$

Then do the sum: 4.35 × 6 = 26.10 or 26.1

Example

0.45 ÷ 5

Put the decimal points underneath each other.

$$\begin{array}{r} 0.09 \\ 5\overline{)0.45} \end{array}$$ 0.45 ÷ 5 = 0.09

Exercise 1

Now you work these out the same way:

1. 2.5 × 3	**4.** 3.66 ÷ 6	**7.** 205.15 ÷ 5
2. 3.75 × 4	**5.** 19.5 ÷ 3	**8.** 318.6 ÷ 9
3. 1.25 ÷ 5	**6.** 2.25 × 8	**9.** 0.66 × 12

10. Jackie buys 3.5 metres of ribbon costing 16p a metre. How much does it cost altogether?

11. Dermot has four 1.5 kg weights. How much do they weigh altogether?

12. Maxine has 5 packets of biscuits. Each packet weighs 118.5 g. How much do they weigh altogether?

13. A bar of chocolate has 6 segments. The whole bar weighs 0.18 kg. How much does each segment weigh?

14. Adrian has a piece of string 1.62 m long. He cuts it into 9 equal pieces. How long is each piece?

15. Alice has a packet of biscuits weighing 21.6 g. There are 8 biscuits in the packet. How much does each one weigh?

How many?

12 divided by 4 means 'how many 4s in 12?'
The answer is 3. There are 3 4s in 12.

In the same way, $\frac{1}{2} \div \frac{1}{4}$ means 'how many quarters in a half?'

The answer is 2.

There are 2 quarters in a half.

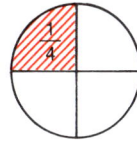

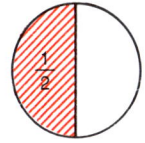

Exercise 1

Work these out the same way. The first one is done for you.

1. $\frac{1}{3} \div \frac{1}{9} = 3$

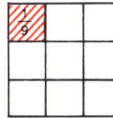

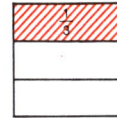

There are 3 ninths in one third

2. $\frac{1}{2} \div \frac{1}{6}$ **4.** $\frac{1}{5} \div \frac{1}{10}$ **6.** $\frac{2}{3} \div \frac{1}{6}$

3. $\frac{1}{4} \div \frac{1}{8}$ **5.** $\frac{3}{4} \div \frac{1}{8}$ **7.** $\frac{3}{4} \div \frac{1}{12}$

We can also use equivalent fractions to divide.

$$\frac{3}{4} \div \frac{1}{3}$$

We use equivalent fractions to convert $\frac{3}{4}$ and $\frac{1}{3}$ to twelfths.

$$\frac{3}{4} \xrightarrow{\times 3} = \frac{9}{12} \qquad \frac{1}{3} \xrightarrow{\times 4} = \frac{4}{12}$$

So we have $\frac{9}{12} \div \frac{4}{12}$

Because both fractions are the same sort of share (both twelfths), this is the same as $9 \div 4$.

We can rewrite $9 \div 4$ as $\frac{9}{4}$ nine quarters

This is the same as $2\frac{1}{4}$. There are $2\frac{1}{4}$ thirds in 3 quarters.

Exercise 2

Work these out using equivalent fractions.
The first one has been done for you.

1. $\dfrac{1}{2} \div \dfrac{1}{5}$ $\overset{\times 5}{\dfrac{1}{2} = \dfrac{5}{10}}$ $\overset{\times 2}{\dfrac{1}{5} = \dfrac{2}{10}}$ so we have $\dfrac{5}{10} \div \dfrac{2}{10}$

This is the same as $5 \div 2$, which can be written as $\dfrac{5}{2}$

So $\dfrac{1}{2} \div \dfrac{1}{5} = \dfrac{5}{2} = 2\dfrac{1}{2}$

There are $2\dfrac{1}{2}$ fifths in one half.

Now try these:

2. $\dfrac{1}{4} \div \dfrac{1}{3}$ **6.** $\dfrac{3}{5} \div \dfrac{9}{10}$ **10.** $\dfrac{5}{9} \div \dfrac{1}{3}$

3. $\dfrac{1}{2} \div \dfrac{1}{3}$ **7.** $\dfrac{3}{8} \div \dfrac{5}{12}$ **11.** $\dfrac{7}{10} \div \dfrac{1}{4}$

4. $\dfrac{2}{3} \div \dfrac{1}{6}$ **8.** $\dfrac{1}{3} \div \dfrac{3}{5}$ **12.** $\dfrac{3}{4} \div \dfrac{5}{8}$

5. $\dfrac{3}{4} \div \dfrac{2}{3}$ **9.** $\dfrac{3}{5} \div \dfrac{1}{3}$ **13.** $\dfrac{1}{4} \div \dfrac{4}{5}$

Exercise 3

Here are some written questions:

1. Jackie has a cake. She eats $\dfrac{1}{5}$ of it and gives $\dfrac{2}{5}$ to Maxine. How much is left?

2. Michael is making biscuits. He uses $\dfrac{1}{2}$ lb flour, $\dfrac{3}{8}$ lb butter and $\dfrac{1}{4}$ lb sugar. How much do the ingredients weigh altogether?

3. Maria works in a cycle shop. Speedo racing cycles are half price in the sale. The original price was £226.50. How much do they cost now?

4. Patricia has a cake cut into 20 slices.
She gives $\frac{1}{4}$ to Roy. How many slices is this?

She gives $\frac{1}{5}$ to Gulsen. How many slices is this?

She gives $\frac{1}{10}$ to Alice. How many slices is this?

How many slices are left?

5. Darren has £3 in his pocket.
He gives $\frac{2}{3}$ to Roy. How much is this?

He gives $\frac{1}{6}$ to Patricia. How much is this?

How much does Darren have left?
What fraction of £3 is this?

6. Maxine buys a dress half price in a sale.
It originally cost £19.50. How much does Maxine pay?

7. Dermot is making pastry. He uses $\frac{1}{2}$ lb flour, $\frac{1}{4}$ lb fat and
$\frac{1}{8}$ lb sugar. How much do the ingredients weigh altogether?

8. There are 24 pupils in 4G

$\frac{1}{4}$ like playing tennis

$\frac{1}{3}$ like playing basketball

$\frac{1}{8}$ like running

$\frac{1}{12}$ like trampolining

How many pupils like each sport?
How many pupils do not like any of these sports?

Areas

The **area** of a shape is the amount of space it takes up.
We use squares to measure area because squares fit together easily.

In this section we will use centimetre squares. Each side is one centimetre long.

Each square is one square centimetre.
We write one square centimetre 1 cm^2.

Look at the rectangle below. It has been divided into centimetre squares.

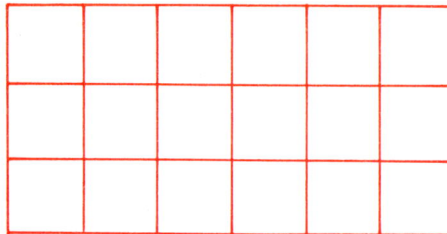

There are three rows of squares. Each row has six squares in it.

So there are 3 × 6 = 18 squares in the whole shape.
The area of the rectangle is 12 cm^2.

Exercise 1

Look at these rectangles.
Work out their areas the same way.

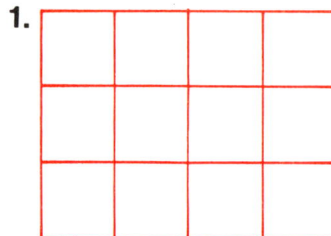

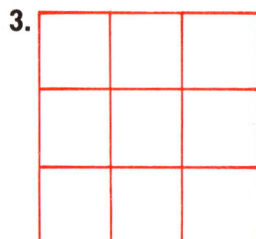

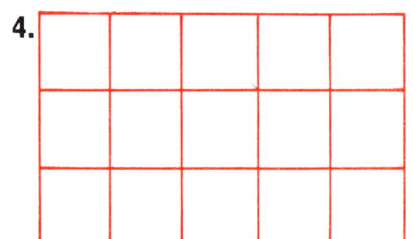

1.

2.

3.

4.

96

5.

6.

7.

8.

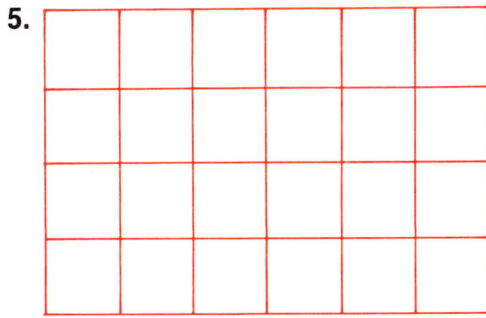

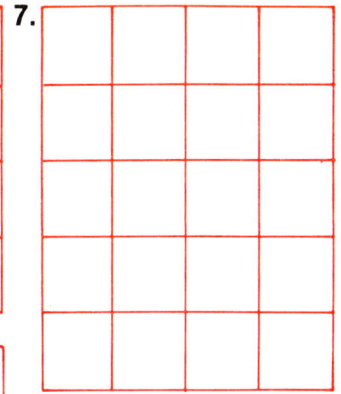

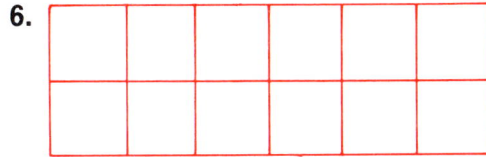

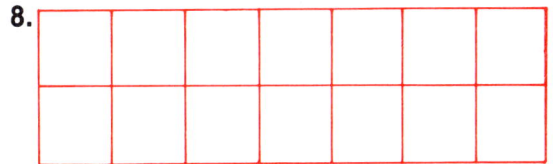

Now work these out.
The first one has been done for you.

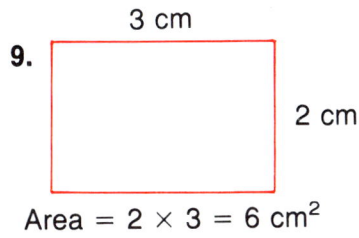

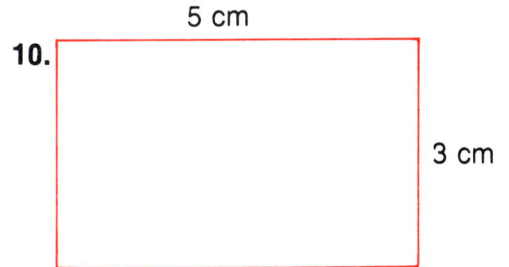

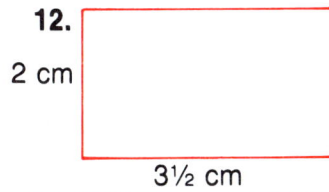

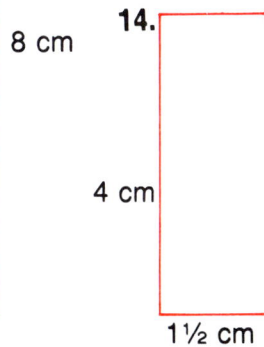

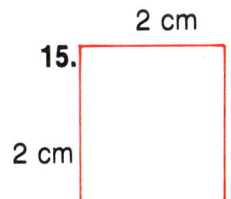

9. 3 cm

2 cm

Area = 2 × 3 = 6 cm²

10. 5 cm

3 cm

11.

8 cm

3 cm

12.

2 cm

13. 3½ cm

1 cm

6 cm

14.

4 cm

1½ cm

15. 2 cm

2 cm

Grams and kilograms

We use **grams (g)** and **kilograms (kg)** for weighing things.

Grams are used for weighing very light things.
A marble weighs about 5 g.
Kilograms are used for weighing heavier things.
Sugar is usually sold in 1 kg bags.
Sometimes we say 'kilo' for kilogram.

kilo means thousand.
So there are a thousand grams in a **kilo**gram.

1000 g = 1 kg

We can use columns to convert grams to kilograms and back again

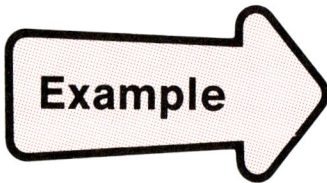

Example

we don't usually use these

kg	hg	dag	g
1	0	0	0
4	7	0	0
3	2	1	4
0	3	5	0

1 kg = 1000 g
4.7 kg = 4700 g
3124 g = 3.214 kg
350 g = 0.35 kg

we write in a zero →

Exercise 1

Copy these and fill in the gaps.

1. 2000 g = ——— kg
2. 2500 g = ——— kg
3. 6000 g = ——— kg
4. 4780 g = ——— kg
5. 215 g = ——— kg
6. 120 g = ——— kg
7. 35 g = ——— kg
8. 280 g = ——— kg

9. 3.5 kg = ——— g
10. 4.2 kg = ——— g
11. 6.305 kg = ——— g
12. 0.5 kg = ——— g
13. 0.06 kg = ——— g
14. 0.543 kg = ——— g
15. 0.018 kg = ——— g
16. 36.342 kg = ——— g

Missing numbers

Finding the missing number in an equation is called **solving** the equation.

We write letters to stand for the missing numbers.

Example

$3 + a = 9$ Find the value of a.

This is like writing: $3 + \square = 9$

We fill in the box: $3 + \boxed{6} = 9$

So $a = 6$

Example

Solve $4p = 20$ **Remember:** $4p = 4 \times p$

This is like writing: $4 \times \square = 20$

We fill in the box: $4 \times \boxed{5} = 20$

So $p = 5$

Shorthand reminders:

$2p = 2 \times p$ $\dfrac{p}{2} = p \div 2$

Exercise 1

Solve these equations (find the value of the letter).

1. $3 + a = 7$
2. $2 - a = 1$
3. $4 + p = 6$
4. $4r = 12$
5. $5b = 10$
6. $3p = 9$
7. $6 + p = 8$
8. $p + 4 = 16$
9. $r - 3 = 10$
10. $3 + r = 56$
11. $\dfrac{r}{5} = 4$

12. $3m = 18$
13. $\dfrac{m}{6} = 5$
14. $2b = 36$
15. $4p = 48$
16. $\dfrac{36}{m} = 3$
17. $100r = 300$
18. $6m = 72$
19. $\dfrac{84}{a} = 12$
20. $p - 16 = 8$

Calculator Problems

Calculators make it much easier to work out the answers to problems.
We work out which sum to do.
Then we do the sum on the calculator.

Example

There are 192 pupils in the 4th year.
They must be put into 8 equal sized classes.
How many pupils will there be in each class?

We need to do a division sum

$192 \div 8 = 24$

There will be 24 pupils in each class.

Exercise 1

Find the answers to these problems.
Work out which sum to do.
Then do the sum on the calculator.

1. Curtis goes shopping.
He buys a coat for £33.50, a jumper for £8.99 and a pair of jeans for £9.85.
How much does he spend altogether?
How much change does he get from £55?

2. Maxine is building a wall.
She uses 35 layers of bricks.
Each layer consists of 76 bricks.
How many bricks does she use altogether?

3. A milkman delivers 54 crates of milk.
Each crate contains 24 bottles.
How many bottles does he deliver?

4. Paulette earns £4316 a year.
How much does she earn a week?

5. Barjinder earns £78.50 per week.
How much does he earn a year?

6. Which job pays more and by how much?
Job A: £4992 per year
Job B: £94 per week

7. A housing estate is going to be built on a piece of land.
The area of the land is 4250 m^2.
Each house needs a plot of land area 85 m^2.
How many houses can be built?

8. A shopkeeper sells coats.
They cost him £15 each.
He sells them for £27.35 each.
How much profit does he make on each one?
He sells 26 coats. How much profit does he make altogether?

9. Here is the price list for a café:

cod and chips	**£1.35**
egg and chips	**£1.15**
omelet and chips	**95p**
apple pie	**47p**
ice cream	**50p**
coffee	**25p**
tea	**20p**
milk	**35p**

How much will these meals cost?

(a) egg and chips, apple pie, milk
(b) cod and chips for two people, one tea, one coffee
(c) omelet and chips, egg and chips, two apple pies, two teas
(d) cod and chips, two lots of egg and chips, one apple pie, two ice creams, three glasses of milk
(e) two apple pies, four ice creams, three coffees, three teas

Co~ordinates

We use **co-ordinates** to say where things are on maps.

Look at the map below.

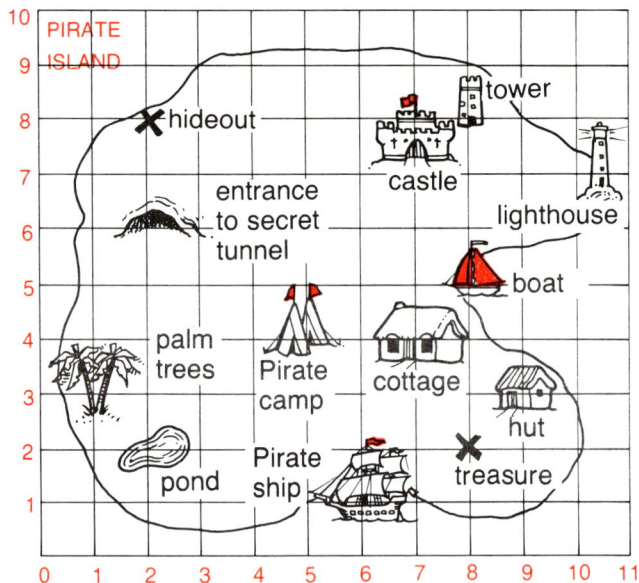

The boat is at (8,5).
It is 8 along and 5 up.   We say the co-ordinates of the boat are
(8,5).

The castle is at $(7,7\frac{1}{2})$.
It is 7 along and $7\frac{1}{2}$ up.   We say the co-ordinates of the castle
are $(7,7\frac{1}{2})$.

Exercise 1

A.   **1.** What is at these points:

    (a) (2,2)?   **(c)** (8,8)?   **(e)** (2,6)?

    (b) (6,1)?   **(d)** (5,4)?   **(f)** $(10\frac{1}{2},6\frac{1}{2})$?

   2. What are the co-ordinates of:

    (a) the hideout?   **(c)** the palm trees? **(e)** the hut?

    (b) the treasure? **(d)** the cottage?   **(f)** the castle?

B. Use this map to answer the questions below.

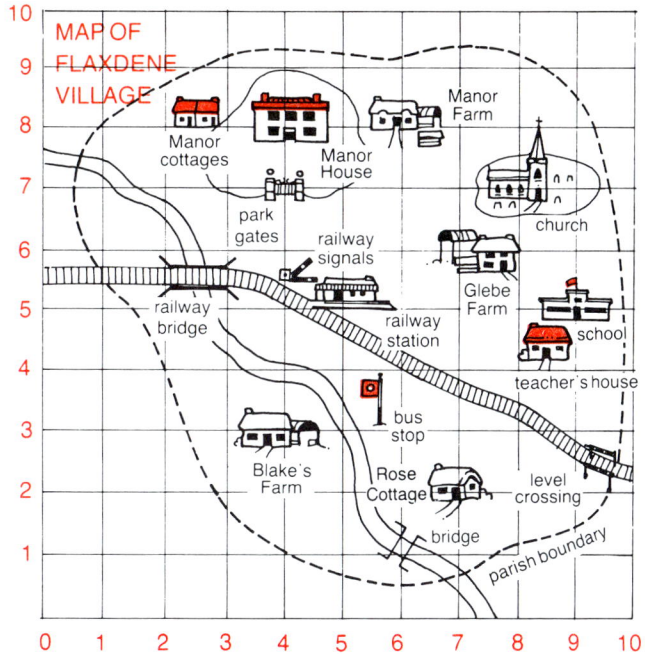

1. What is at:

(a) (6,1)?

(b) (6,8)?

(c) (2½,8)?

(d) (7,2)?

(e) (5,5)?

(f) (7,6)?

(g) (5½,3½)?

(h) (4,5½)?

(i) (8½,4½)?

2. What are the co-ordinates of:

(a) Blake's Farm?

(b) Manor House?

(c) park gates?

(d) school?

(e) Manor Cottage?

(f) railway bridge?

(g) teacher's house?

(h) level crossing?

(i) church?

Pictograms

Pictograms are a way of showing data using pictures. Here is an example.

year	number of cars sold by Driveaway Motors
1985	🚗 🚗 🚗 🚗
1984	🚗 🚗 🚗 🚗
1983	🚗 🚗 🚗 🚗 🚗 🚗 🚗
1982	🚗 🚗 🚗 🚗 🚗
1981	🚗 🚗 🚗
1980	🚗 🚗 🚗 🚗 represents 100 cars

How many cars do you think 🚗 represents?

Exercise 1

Use the pictogram to answer these questions:

1. How many cars did Driveaway Motors sell in 1984?

2. How many did they sell in 1981?

3. How many more cars did they sell in 1983 than in 1982?

4. How many fewer cars did they sell in 1980 than in 1985?

5. How many more cars did they sell in 1985 than in 1981?

When you draw a pictogram, remember:

give it a title
make every picture the same size
space the pictures evenly
say what the picture represents.

You will find it easier if you use squared paper.

Exercise 2

1. Here is a table showing the number of people visiting Blogg's Diner each day one week:

day	no. of people
Monday	40
Tuesday	20
Wednesday	30
Thursday	25
Friday	50
Saturday	65

Draw a pictogram to show this data.
Use one symbol to represent 10 people.

2. Here is a pictogram showing the number of men and women working for a firm between 1980 and 1985.

year	number of men and women working for the firm
1980	🧍🧍🧍🧍🧍🧍🧍🧍🧍🧍
1981	🧍🧍🧍🧍🧍🧍🧍🧍🧍
1982	🧍🧍🧍🧍🧍🧍🧍
1983	🧍🧍🧍🧍🧍🧍🧍🧍🧍
1984	🧍🧍🧍🧍🧍🧍
1985	🧍🧍🧍🧍🧍🧍🧍🧍

🧍 rep. 10 men   🧍 rep. 10 women

(a) How many men worked for the firm in 1980?
(b) How many women worked for the firm in 1982?
(c) How many women worked for the firm in 1983?
(d) How many people altogether worked for the firm in 1985?
(e) How many people left the firm between 1983 and 1984?
(f) How many people were taken on between 1984 and 1985?

3. Here is a table showing sales of boxes of Pull-it Christmas crackers over five years.

Draw a pictogram to show this data.

Use one symbol to represent 20 000 boxes.

year	no. sold
1981	20 000
1982	60 000
1983	80 000
1984	90 000
1985	120 000

Rule of zeros

Multiply these on the calculator.

1. 20×3 **2.** 50×70 **3.** 50×40 **4.** 20×300
5. 20×30 **6.** 500×6 **7.** 40×40 **8.** 40×400

What do you notice?

When numbers end in zero, we can multiply them very easily.

Take off the zeros
Multiply the numbers
Add on the same number of zeros
as you took off altogether.

This is called the
rule of zeros.

Example

20×3

remove zero: $2 \times 3 = 6$
add zero to answer: $= 60$ —— 1 zero added

$20 \times 30 = 60$

Example

50×40

remove zeros: $5 \times 4 = 20$
add zeros to answer: $= 2000$ —— 2 zeros added

$50 \times 40 = 2000$

Exercise 1

Use the rule of zeros to work these out without a calculator.

1. 20×4
2. 30×50
3. 40×200
4. 30×20
5. 60×60
6. 60×600
7. 10×30
8. 40×4
9. 6×700
10. 40×300

11. 50×500
12. 20×300
13. 2000×50
14. 30×50
15. 120×4
16. 110×50
17. 30×70
18. 80×120
19. 90×40
20. 600×1100

We can use the rule of zeros to help us check answers to calculations.

Sharon has to do the sum 43 × 5.
She gets the answer 215.

43 is about 40
So 43 × 5 is about 40 × 5.

40 × 5 = 200.

215 is about 200.
So Sharon's answer is probably correct.

Sharon gets the answer 836 for the sum 220 × 38.

220 is about 200
38 is about 40.
So 220 × 38 is about 200 × 40.

200 × 40 = 8000.

836 is nowhere near 8000.
So Sharon's answer is probably not correct.

Exercise 2

Use the rule of zeros to check these calculations.

1. 43 × 3 = 129
2. 54 × 42 = 2268
3. 120 × 35 = 420
4. 460 × 18 = 8280
5. 230 × 120 = 276 000
6. 63 × 49 = 308
7. 48 × 72 = 3456
8. 108 × 470 = 50 760
9. 390 × 210 = 81 900
10. 46 × 87 = 4002

11. 63 × 15 = 945
12. 79 × 25 = 197
13. 402 × 191 = 767 820
14. 360 × 78 = 28 080
15. 197 × 212 = 417 644
16. 490 × 11 = 5390
17. 680 × 29 = 19 720
18. 370 × 20 = 740
19. 69 × 433 = 298 770
20. 562 × 17 = 9554

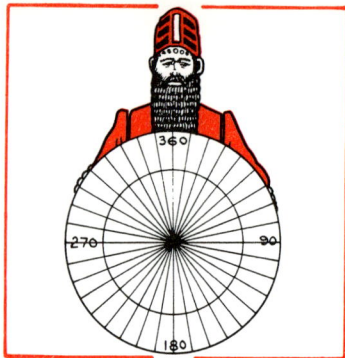

Degrees

Angles are measured in **degrees**.

The Babylonians started measuring in degrees.
They chose 360 degrees for one full turn because 360 has lots of factors.

One full turn is 360 degrees  360 degrees

So one half turn is 180 degrees 180 degrees

One quarter turn is 90 degrees 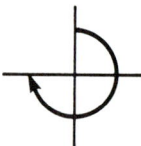 90 degrees

One three quarter turn is 270 degrees 270 degrees

We use a little circle ° to mean 'degrees'.

We can estimate the size of an angle by seeing about how much of a turn it is.

Example

This angle is a bit less than a quarter turn.

So it is a bit less than 90°.

Estimate: 80°.

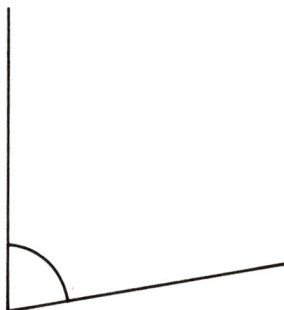

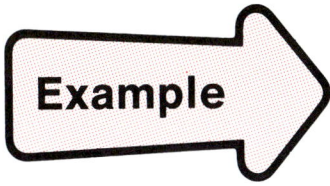

This angle is more than 90° but less than 180°.

It is nearer 180° than 90°.

Estimate: 160°.

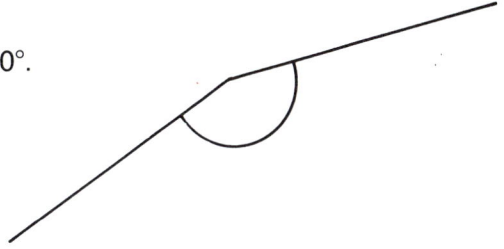

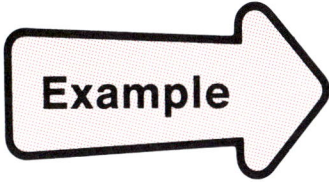

This angle is more than 270° but less than 360°.

It is nearer 270° than 360°.

Estimate: 290°.

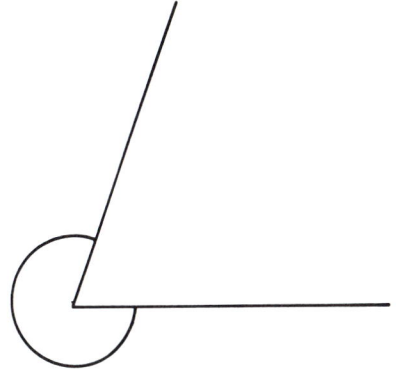

Exercise 1

A. Look at the angles below.

Estimate the size of each one and match it to one of these angles:

1. 80°   **2.** 110°   **3.** 55°   **4.** 170°   **5.** 185°
6. 20°   **7.** 260°   **8.** 340°   **9.** 355°   **10.** 135°

(a)
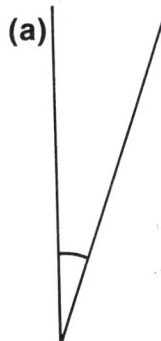

(b)

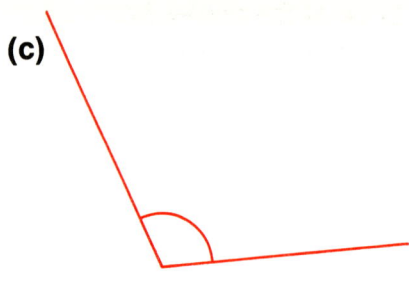

(c)

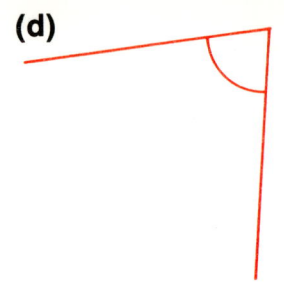

(d)

(e)

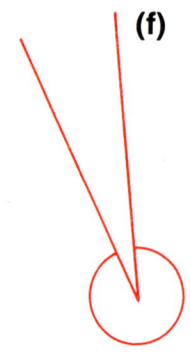

(f)

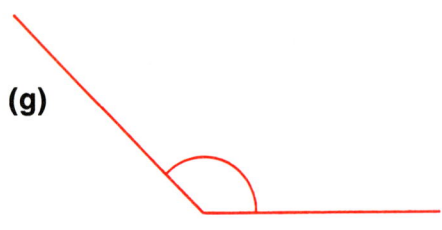

(g)

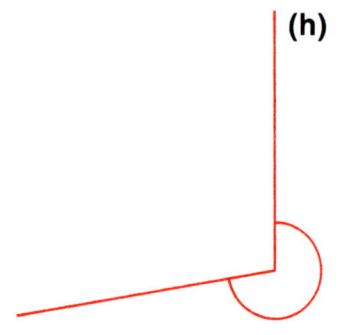

(h)

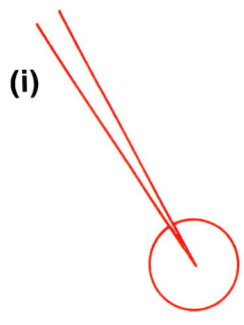

(i)

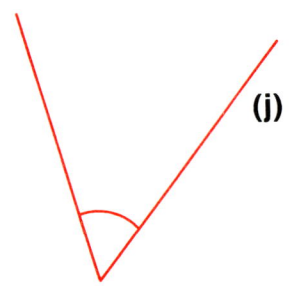

(j)

B. Look at the angles below.
Estimate the size of each one.

1.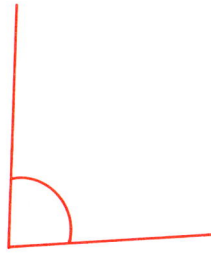

2.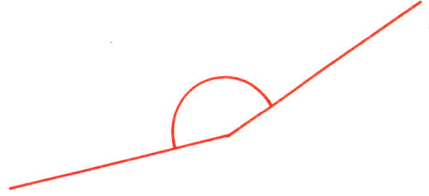

3.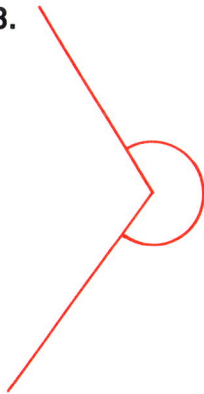

4.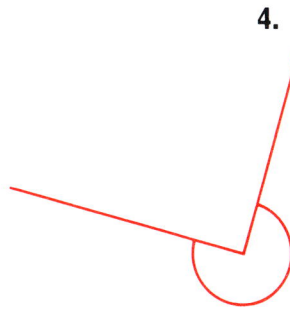

5.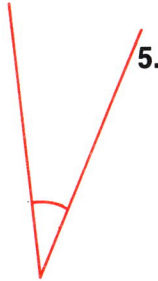

6.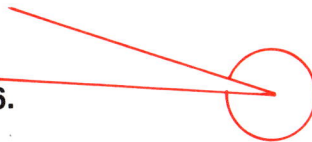

7.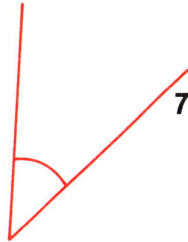

8.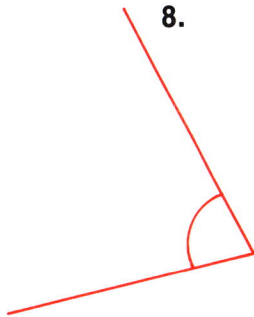

9.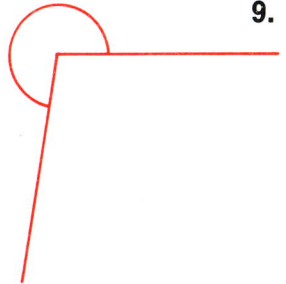

10.

Parallelogram areas

It does make a rectangle

Cut a parallelogram out of paper.

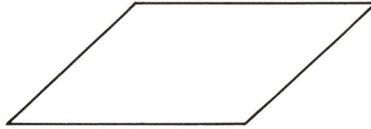

Cut a triangle off one end as in the diagram below:

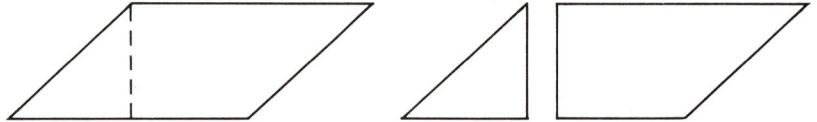

Fit the triangle onto the other side of the parallelogram to make a rectangle:

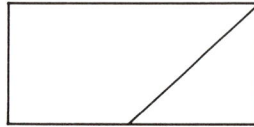

Do the same thing with three different parallelograms.
Stick the rectangles in your book.

You can always cut the end off a parallelogram like this and make a rectangle.

So the area of the parallelogram is the same as the area of the rectangle.

The area of a rectangle is length × width.

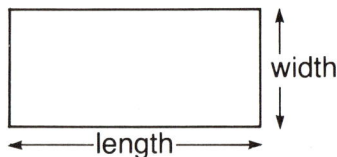

The area of a parallelogram is base × height.

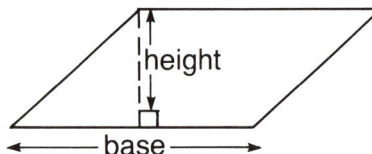

The height is always at right angles to the base.

Exercise 1

Work out the areas of these parallelograms.
The first one has been done for you.

1.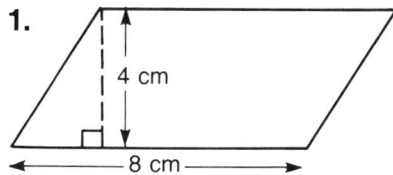

Area of a parallelogram
= base × height
= 8 × 4
= 32 cm²

4 cm
8 cm

2.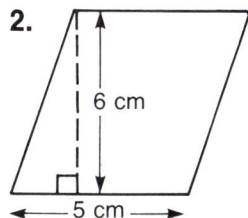

6 cm
5 cm

3.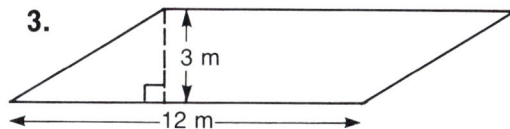

3 m
12 m

4.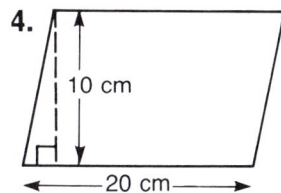

10 cm
20 cm

5.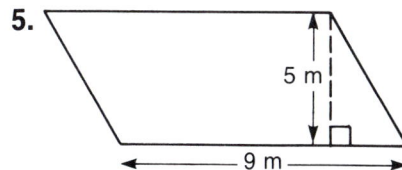

5 m
9 m

6.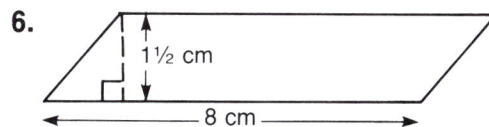

1½ cm
8 cm

7.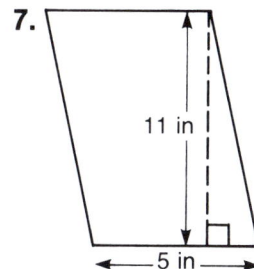

11 in
5 in

8.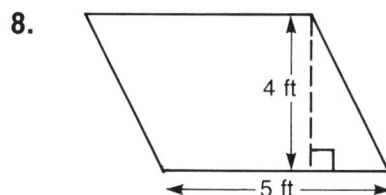

4 ft
5 ft

9.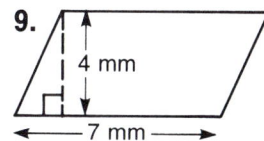

4 mm
7 mm

10.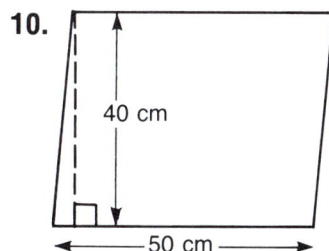

40 cm
50 cm

11.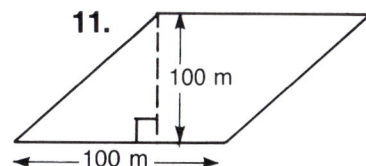

100 m
100 m

Check-up 5

1. Work these out:

(a) 2.6×4 **(b)** 43.5×6 **(c)** $27.3 \div 3$ **(d)** $4.86 \div 6$

2. Work these out:

(a) $\dfrac{1}{2} \div \dfrac{1}{4}$ **(b)** $\dfrac{3}{4} \div \dfrac{5}{8}$ **(c)** $\dfrac{2}{3} \div \dfrac{1}{6}$

3. Work out the areas of these rectangles:

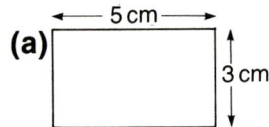

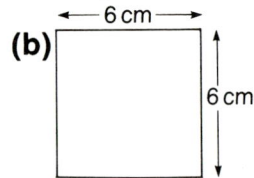

(a) 5 cm, 3 cm

(b) 6 cm, 6 cm

4. Copy these and fill in the gaps:

(a) $3500 \text{ g} = \underline{\hspace{2cm}} \text{ kg}$ **(b)** $3.64 \text{ kg} = \underline{\hspace{2cm}} \text{ g}$

5. Solve these equations:

(a) $5 + a = 7$ **(b)** $4p = 28$ **(c)** $\dfrac{r}{6} = 3$

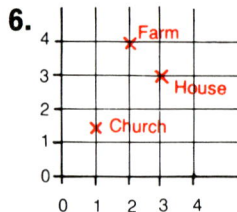

6.

Farm, House, Church coordinate grid

(a) What is at (2,4)?

(b) What are the coordinates of:

(i) the house?

(ii) the church?

7. Here is a chart showing the number of rainy days over six months.

Draw a pictogram to show this data
Use one symbol to represent 2 rainy days.

month	no.	month	no.
January	12	April	14
February	18	May	10
March	16	June	7

8. Work these out without a calculator:

(a) 20×30 **(b)** 4×50 **(c)** 60×300

9. Estimate the size of these angles:

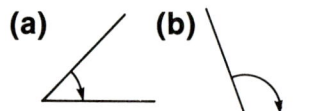

(a)   **(b)**

10. Work out the areas of these parallelograms:

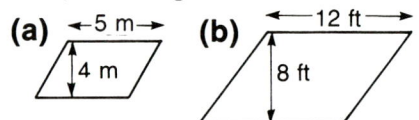

(a) 5 m, 4 m   **(b)** 12 ft, 8 ft

I've never bought soap in my life

Out of a hundred

Eva and her friends do a survey. They interview 100 people.
Here are their results:

43 people used 'Bubblo' soap
36 people used 'Sudzo' soap
21 people used other brands of soap

'Per cent' is another way of saying 'out of 100'.

We write **%**.

43 people out of 100 people used 'Bubblo' soap.
We say 43% of people used 'Bubblo' soap.

Exercise 1

Copy and complete these sentences:

1. _____ % of people used 'Sudzo' soap.

2. _____ % of people used other brands of soap.

Darren does a survey with his friends.
They also interview 100 people.
They write their results as percentages:

36% play basketball
27% play volleyball
16% play snooker
12% play tennis
 9% play table tennis

'Per cent' means 'out of 100'.
36 people out of 100 people play basketball.
Darren's friends interviewed 100 people.
So 36 of them play basketball.

Exercise 2

1. How many people played the other games?

The fourth year held elections for School Council representatives.
200 pupils voted.

Maxine writes the results as percentages.

Ilknur	43%
Dermot	22%
Ovid	13%
Alice	12%
Patricia	7%
Sanjeev	3%

Sanjeev got 3% of the votes.
3 people out of every hundred voted for him.
200 people voted.
So the number of people who voted for Sanjeev is $2 \times 3 = 6$.

Exercise 3

1. How many people voted for each of the other candidates?

2. Here are the results of a survey of favourite drinks.
500 people were surveyed.

Coffee	18%
Tea	27%
Milk	10%
Lemonade	8%
Fruit juice	12%
Hot chocolate	19%
Coke	6%
Total	100%

18 out of every 100 liked coffee best.
500 people were surveyed.
So $5 \times 18 = 90$ people liked coffee best.

How many people liked the other drinks best?

You can check your answers by adding them to see if there
are 500 people altogether.

3. Here are the results of a survey of favourite sorts of television programmes. 300 people were surveyed.

Comedy series	35%
Soap operas	21%
Plays	2%
Music programmes	19%
Documentaries	4%
Magazine programmes	12%
News	7%
Total:	100%

How many people liked each sort of programme best?

Zahida surveys 200 people.
She asks them what sort of books they like reading most.

Here are her results:

type of book	number of people
Thrillers	62
Ghost stories	24
Romances	38
Biographies	12
Detective stories	46
Historical novels	18
Total:	200

Zahida wants to write her results as percentages.

62 people out of 200 like thrillers.
So 62 ÷ 2 = 31 people out of 100 like thrillers.
31% of the people surveyed like thrillers.

Exercise 4

1. Change the rest of Zahida's results as percentages.

You can check your answers by seeing if they add up to 100%.

2. Barjinder surveys 300 people.
He asks them what their favourite food is.
Here are the results:

favourite food	number of people
fish and chips	81
chocolate cake	96
curry	45
kebabs	15
roast beef	54
vegetable stew	9
Total:	300

Convert Barjinder's results to percentages.

3. Maxine's teacher sets the class a test.
It is marked out of 50.
Here are some of the results:

pupil	marks out of 50
Maxine	42
Darren	38
Dermot	40
Ilknur	36
Robert	27
Paulette	24
Simon	15

Maxine's teacher converts the results to percentages.

Maxine got 42 marks out of 50.
So she got $42 \times 2 = 84$ out of 100.
Maxine got 84% in the test.

Work out the percentage scores for the other pupils.

4. 70% of the 5th year like watching football.

(a) What percentage do not like watching football?
(b) There are 200 pupils in the 5th year.
How many like watching football?
(c) How many do not like watching football?

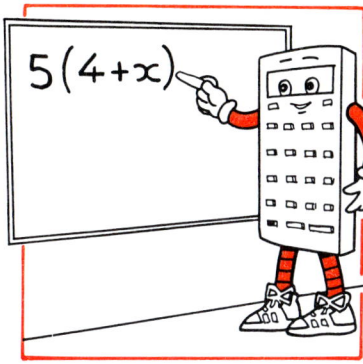

Brackets

Do these sums in your head, then on the calculator.

① $2 + 4 \times 3$

② $3 \times 6 + 2$

③ $4 + 2 \times 6 - 3$

④ $4 + 6 \div 2 + 3$

You may get two different answers for each sum:

① 14 or 18   ② 20 or 24   ③ 13 or 18   ④ 10 or 2.

Your answers depend on which part of the sum you do first.

Scientific calculators always do calculations in the same order.
They multiply or divide first, then add or subtract.
The calculator rewrites the sum before doing it, then does the parts in order.

① is rewritten 4×3 then $+ 2$

② is rewritten 3×6 then $+ 2$

③ is rewritten 2×6 then $+ 4 - 3$

④ is rewritten $6 \div 2$ then $+ 4 + 3$

Exercise 1

Rewrite these sums in the order the scientific calculator will do them.
Then do the sum.

1. $3 + 4 \times 2$

2. $10 \times 2 - 3$

3. $4 + 6 \div 3$

4. $3 + 7 \times 4$

5. $20 + 7 \times 2$

6. $4 + 5 \times 3 + 2$

7. $8 \div 4 - 1$

8. $12 + 6 \times 3$

9. $14 + 10 \div 2$

10. $5 + 16 \div 4$

11. $24 + 4 \times 3 - 8$

12. $14 + 12 \div 3 - 5$

We do not usually rewrite sums to show which order to do them in. We use **brackets** instead.
We put brackets round the parts of the sum we want done first.

Here are the first four sums from page 119 again.
They have been rewritten using brackets.

① $(2 + 4) \times 3$
② $3 \times (6 + 2)$
③ $(4 + 2) \times (6 - 3)$
④ $(4 + 6) \div (2 + 3)$

We work out the sums inside the brackets first:

① $(2 + 4) \times 3$

$= 6 \times 3 = 18$

③ $(4 + 2) \times (6 - 3)$

$= 6 \times 3 = 18$

② $3 \times (6 + 2)$

$= 3 \times 8 = 24$

④ $(4 + 6) \div (2 + 3)$

$= 10 \div 5 = 2$

This is how we got the second set of answers on page 119.

Exercise 2

Work these out in the same way.
Do the sums inside the brackets first.

1. $(4 + 3) \times 2$

2. $4 \times (3 + 6)$

3. $20 - (4 \times 3)$

4. $(5 + 4) \times 6$

5. $15 - (2 \times 5)$

6. $(7 \times 2) - 4$

7. $(8 + 2) \times (4 - 3)$

8. $(5 - 3) \times (8 + 1)$

9. $(10 + 2) \div (6 - 3)$

10. $(5 + 15) \div (2 + 3)$

11. $(2 + 7) \times (4 + 5)$

12. $(16 + 4) \div (32 - 12)$

More co-ordinates

We can use **co-ordinates** to describe pictures on grids.

We describe the shape by writing down the co-ordinates of the crosses at the corners.

We write them in the order they are joined up.

It does not matter where we start.

Here are the co-ordinates of the letter on the right starting at the point marked 'START':

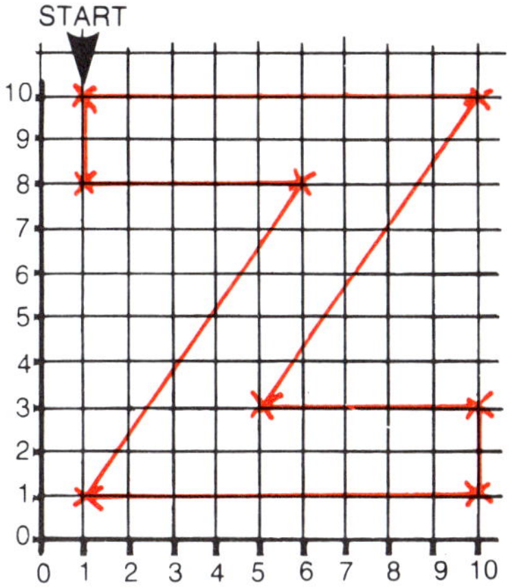

(1,10), (10,10), (5,3), (10,3), (10,1), (1,1), (6,8), (1,8), (1,10)

Exercise 1

Here are some more pictures on grids.
Write down the co-ordinates of the crosses in order.

Always start at the cross marked 'START'.

1.

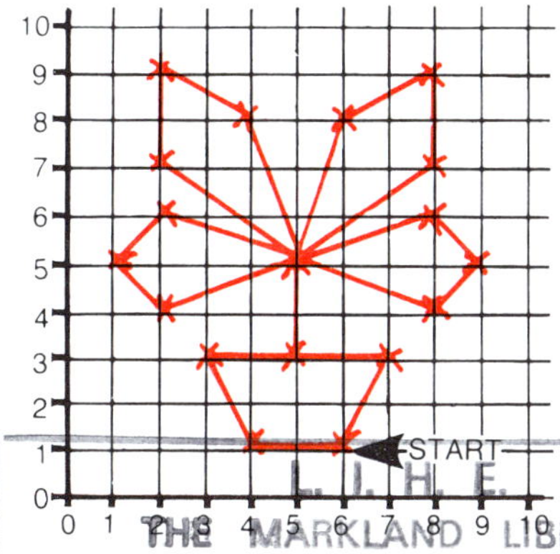

2.

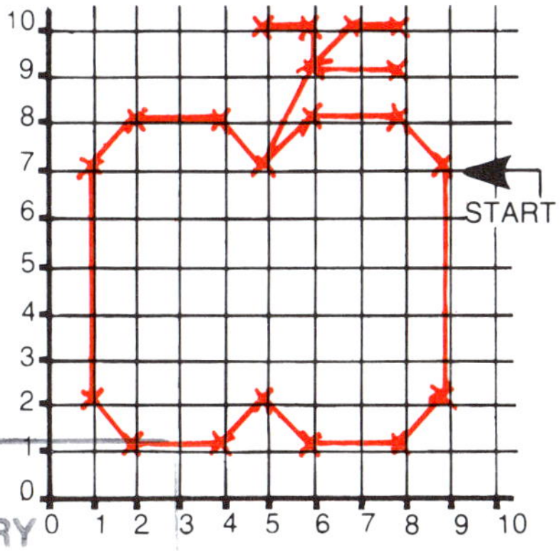

3.

4.

5.

6.

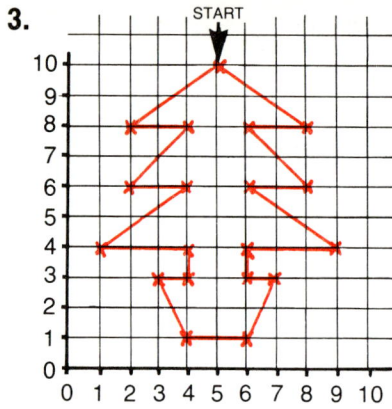

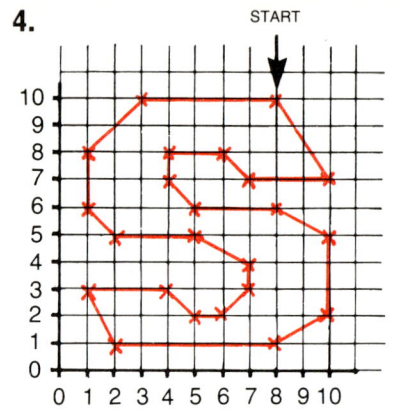

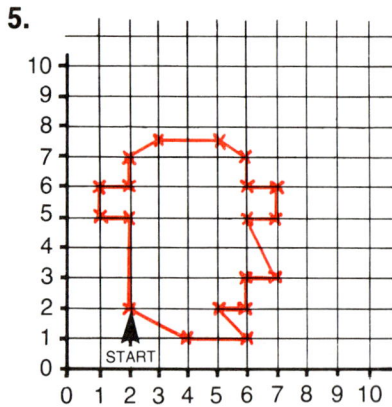

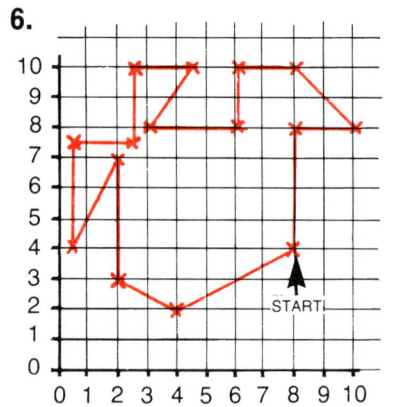

We can make grid pictures from co-ordinates.
We use the co-ordinates to tell us where to mark the crosses.
This is called **plotting points** on the grid.

Example

Draw a grid on centimetre square paper, mark it up to 10 across and up.

Plot these points and join them up in order:

(1,1) (1,7) (3,7) (3,3) (7,3) (7,1) (1,1)

Your picture should start off like this:

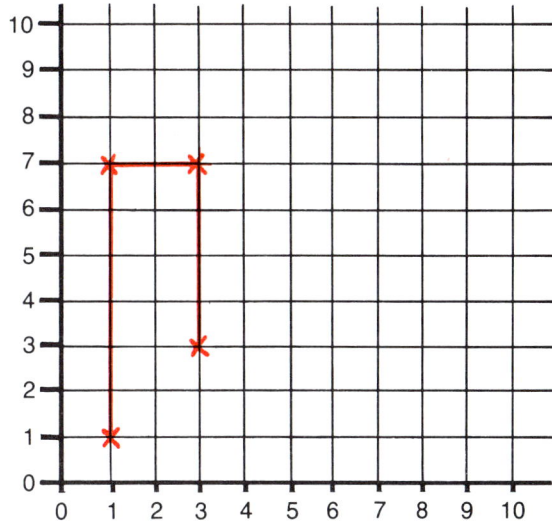

Now you finish it.

Exercise 2

For each question:
Draw a grid.
Mark it up to 10 across and up.
Plot the points and join them up in order.

1. (2,1) (3,1) (5,3) (3,3) (3,6) (1,4) (3,4) (3,3) (1,3) (1,2) (2,1)

2. (5,1) (6,2) (4,8) (5,9) (6,8) (5,5) (6,7) (7,6) (5,5) (8,6) (9,5)
(8,4) (5,5) (7,4) (6,3) (5,5) (4,3) (3,4) (5,5) (2,4) (1,5) (2,6)
(5,5) (3,6) (4,7) (5,5) (4,2) (5,1)

3. (1,4) (2,3) (3,3) (3,2) (4,1) (7,1) (8,2) (10,7) (9,7) (8,5) (8,8)
(7,8½) (6½,8½) (6½,9) (4½,9) (4½,8½) (4,8½) (3,8) (3,7)
(2,7) (1,6) (1,4)

Triangle areas

Here are some paralellograms:

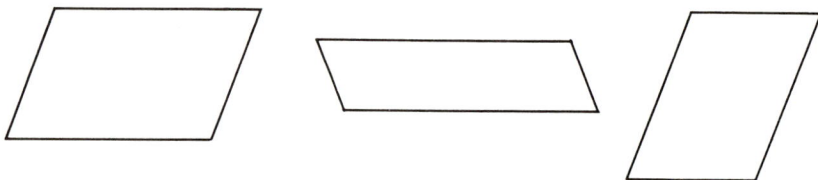

They can all be cut in half diagonally.
Each one makes two triangles.

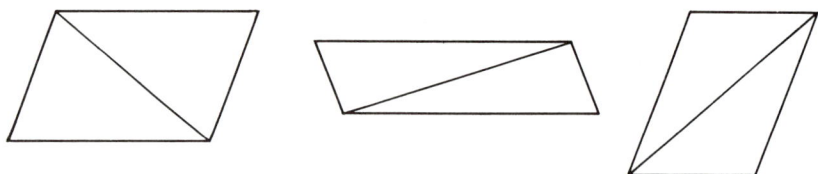

The triangles are both the same size.
So each one is half the size of the parallelogram.
The area of each triangle is half the area of the parallelogram.

The area of a parallelogram is base × height.

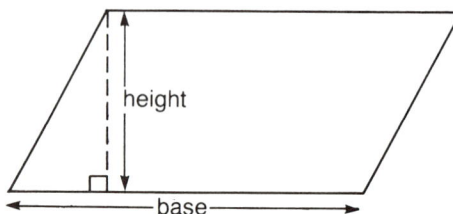

So the area of a triangle is $\frac{1}{2}$ base × height.

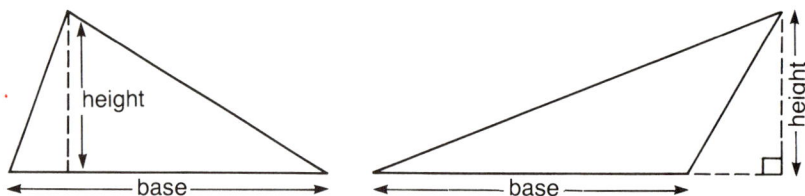

It doesn't matter which side is the base but the height is always
at right angles to it.

Exercise 1

Work out the areas of these triangles.
The first one has been done for you.

Area of a triangle = $\dfrac{1}{2}$ base × height

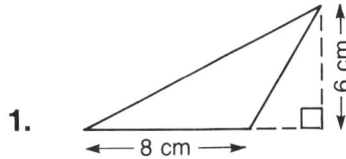

1.

$= \dfrac{1}{2}\ 8 \times 6$

$= \dfrac{1}{2}$ of 48 = 24 cm^2

2.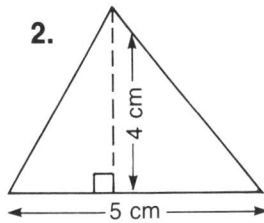

3.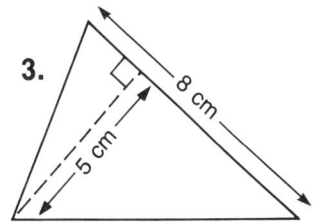

4.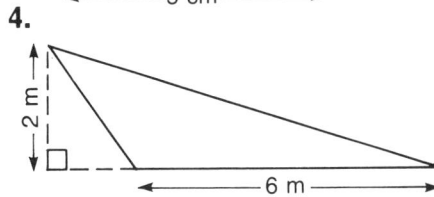

5.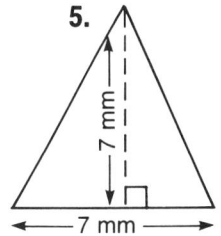

6.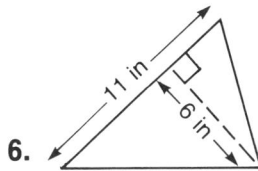

7.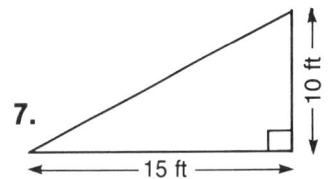

8.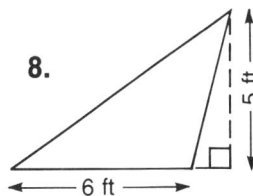

9.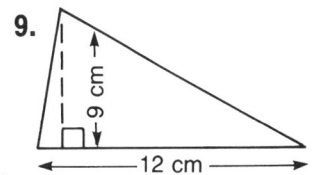

10.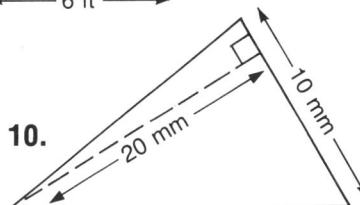

11.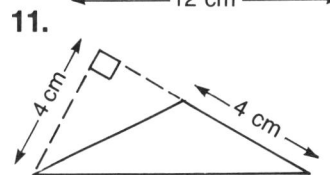

Millilitres and litres

We use **Millilitres (ml)** and **Litres (l)** for measuring liquids.
We use millilitres for very small amounts of liquid.
A teaspoon holds about 5 ml.
Litres are used for larger amounts.
1 l is a bit less than 2 pints.

Sometimes scientists write cc or cm^3 instead of ml.
This is because 1 cubic centimetre of liquid = 1 ml.
1 ml of liquid will fill a cube with sides 1 cm long.

Milli means thousand.
There a a thousand millilitres in a litre.

1000 ml = 1 l.

We can use columns to convert millilitres to litres and back again.

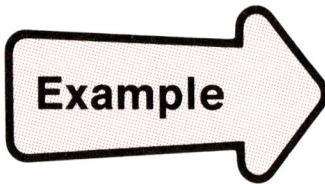

Example

We don't usually use these

l	dl	cl	ml
2 •	0	0	0
0 •	5	2	4
6 •	5	0	0
0 •	4	6	3

we put in a zero →

2 l = 2000 ml

0.524 l = 524 ml

6500 ml = 6.5 l

463 ml = 0.463 l

Exercise 1

Copy these and fill in the gaps:

1.   4000 ml = _____ l
2.   2300 ml = _____ l
3.   450 ml = _____ l
4.   50 ml = _____ l
5.   2196 ml = _____ l
6.   25 ml = _____ l
7. 27 000 ml = _____ l
8. 46 005 ml = _____ l

9.   0.625 l = _____ ml
10. 3.0 l   = _____ ml
11. 0.463 l = _____ ml
12. 2.71 l = _____ ml
13. 0.75 l = _____ ml
14. 36.4 l   = _____ ml
15. 0.05 l = _____ ml
16. 3.75 l = _____ ml

1p in the £1

'1%' means 'one out of a hundred'.

£1 is 100p.
So 1% of £1 is 1p.
25% of £1 is 25p, and so on.

Exercise 1

Work out these amounts:

1. 5% of £1	**5.** 45% of £1	**9.** 47% of £1
2. 20% of £1	**6.** 90% of £1	**10.** 80% of £1
3. 30% of £1	**7.** 75% of £1	**11.** 100% of £1
4. 50% of £1	**8.** 60% of £1	**12.** 15% of £1

15% of £1 is 15p.
So 15% of £3 is 3 × 15p = 45p.

Now work these out:

13. 15% of £2	**17.** 50% of £3	**21.** 60% of 50p (hint: 50p is
14. 4% of £3	**18.** 25% of £4	**22.** 75% of £4 half of £1)
15. 40% of £2	**19.** 10% of £10	**23.** 90% of £5
16. 10% of £5	**20.** 5% of £6	**24.** 30% of £3

25. A dress is sold for £8. In the sale it is reduced by 10%.
How much is it reduced?
What is the new price?

26. Calculators usually cost £12. Jenny gets 25% staff discount.
How much does she pay?

27. Record players cost £100 + 15% V.A.T.
Work out the V.A.T.
How much must the customer pay?

28. A shop takes 25% off everything in the sale.
Work out the new prices of these:

(a) a coat price £16
(b) a bicycle price £60
(c) a scarf price £2
(d) a pack of cards price £1
(e) a radio price £44

How much do I use?

Roy is making pastry in cookery.
The recipe tells him to use twice as much flour as fat.

Roy is using 100 g fat.

So he must use 2×100 g $= 200$ g flour

Exercise 1

How much flour should Roy use for these amounts of fat?

1. 50 g   **4.** 300 g   **7.** 350 g

2. 25 g   **5.** 150 g   **8.** 1000 g

3. 200 g   **6.** 500 g   **9.** 750 g

The teacher says Roy uses half fat to flour.
He uses half as much fat as flour.

Roy uses 400 g flour.

So he must use $400 \div 2 = 200$ g fat.

Exercise 2

How much fat should Roy use for these amounts of flour?

1. 200 g   **4.** 50 g   **7.** 1000 g

2. 600 g   **5.** 100 g   **8.** 800 g

3. 300 g   **6.** 500 g   **9.** 750 g

Michael is making scones.
He uses baking powder.
There is a chart on the side of the tin to show how much to use.

HOW MUCH TO USE

For each 8 oz plain flour:

scones	4 level teaspoons
Victoria sandwich	3 level teaspoons
pastry	1 level teaspoon
fruit cake	2 level teaspoons

Michael uses 16 oz plain flour.
16 oz = 2 × 8 oz
So Michael should use 2 × 4 = 8 level teaspoons of baking powder.

Tricia uses 4 oz plain flour to make her scones.
4 oz = half of 8 oz
So Tricia should use half of 4 = 2 oz of baking powder.

Exercise 3

Work out the amounts of baking powder for these:

1. scones using
   (a) 8 oz flour   **(b)** 24 oz flour   **(c)** 2 oz flour

2. Victoria sandwich using
   (a) 8 oz flour   **(b)** 16 oz flour   **(c)** 4 oz flour

3. pastry using
   (a) 8 oz flour   **(b)** 16 oz flour   **(c)** 4 oz flour

4. fruit cake using
   (a) 4 oz flour   **(b)** 12 oz flour   **(c)** 6 oz flour

Measuring angles

Earlier in this book we estimated the size of angles in degrees. In this section we are going to measure them accurately.

Remember: one full turn   = 360°
so one half turn   = 180°
one quarter turn   = 90°
one three-quarter turn = 270°

We use an angle indicator to measure angles.
Here is a picture of one.

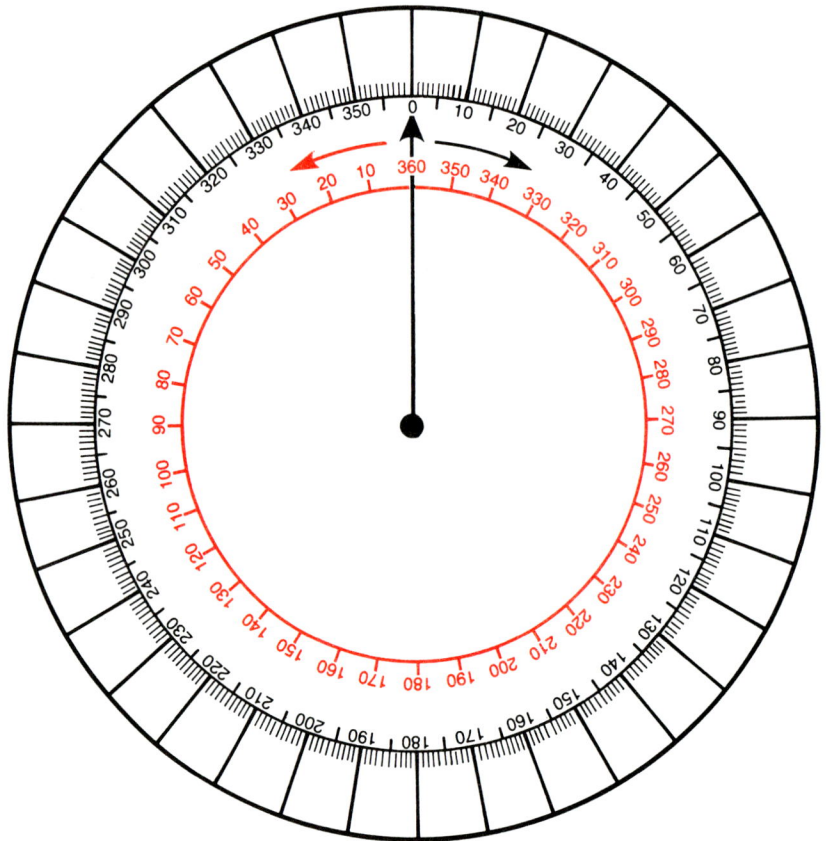

Notice that:

the indicator is made up of two circles
the bigger circle has two scales, clockwise and anticlockwise
each scale is marked in degrees from 0° to 360°
the smaller circle has an arrow from the centre to the edge.

We are going to use the angle indicator to measure this angle:

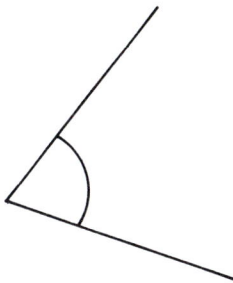

Follow these steps:

1. Make sure the arrow starts pointing at 0°.
2. Line up the arrow with one arm of the angle.
3. Make sure the point of the angle is in the centre of the angle indicator.
4. Turn the smaller circle until the arrow lines up with the other arm of the angle.
5. Count the number of degrees the arrow has turned . . . 10°, 20°, 30° . . . etc.

The angle is 70°. Check you can measure it correctly.

Exercise 1

Now measure these angles.
You may have to turn the book round to read the scale.
Always make sure the arrow starts off pointing at 0°.

1.

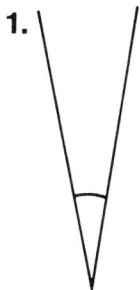

2.

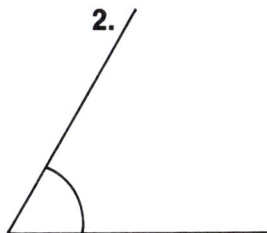

3.

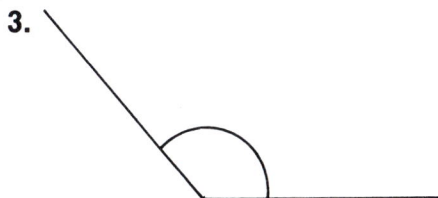

4.

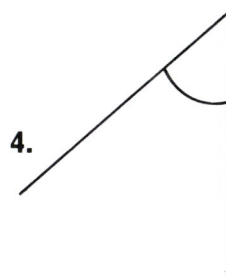

5.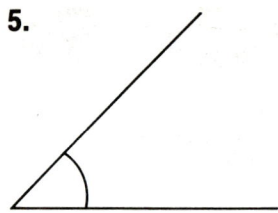

6.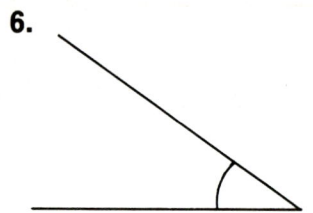

7.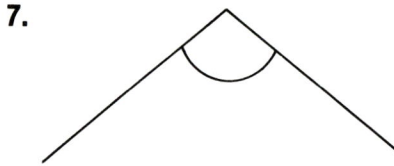

8.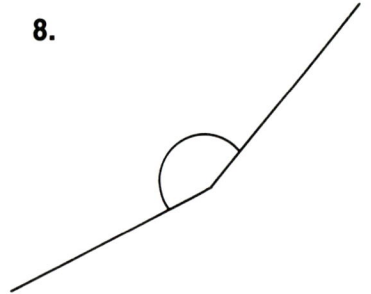

9.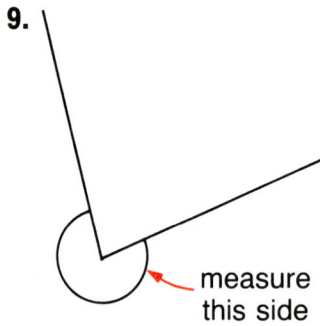

measure this side

10.

11.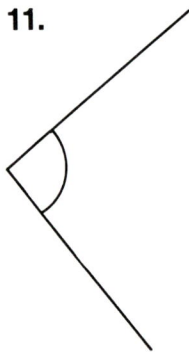

12.

Yards, feet and inches

It's a good thing we have foot rulers for measuring nowadays

When people first started measuring the lengths of things, they used parts of the human body.

A **yard** is about the distance from your nose to your fingertip when your arm is stretched out straight.

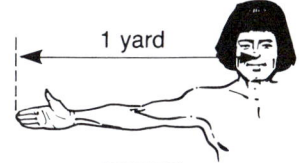

A **foot** is about as long as a large man's foot.

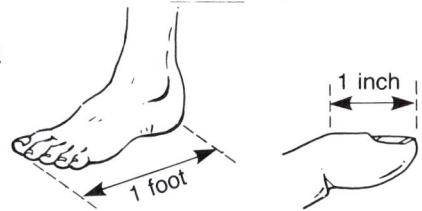

An **inch** is about as long as the end of your thumb.

In Britain we still use yards, feet and inches for measuring many things.
We have made them all standard lengths and do not use bodies any more.

12 inches (in) = 1 foot (ft)   3 feet (ft) = 1 yard (yd)

Exercise 1

1. How many inches in:

(a) 2 ft? (b) 5 ft? (c) 1 yd? (d) 1½ ft? (e) 4 ft? (f) 2 yd?

2. How many feet in:

(a) 3 yd? (b) 2 yd? (c) ½ yd? (d) 4 yd? (e) 10 yd? (f) 5 yd?

3. How many feet is the same as:

(a) 36 in?   (b) 24 in?   (c) 60 in?   (d) 48 in?   (e) 72 in?
(f) 18 in?

4. 3-core electric cable costs 20p per foot. How much will it cost to buy:

(a) 2 ft?   (b) 1 yd?   (c) 3 yd?   (d) 18 in?   (e) 1½ yd?
(f) 7½ yd?

5. Fabric is sold by the yard. How many yards should I buy if I need:

(a) 3 ft? (b) 6 ft? (c) 9 ft? (d) 18 ft? (e) 4 ft 6 in? (f) 18 in?

Check-up 6

1. Here are the results of three pupils in a test:

Convert these results to percentages.

pupil	marks out of 50
Jason	46
Tricia	37
Avril	42

2. Work these out:

(a) $(3 + 2) \times 6$ **(b)** $(4 + 5) \div (6 - 3)$

3. Work out the areas of these triangles:

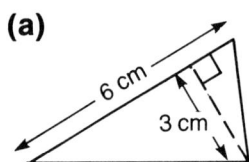

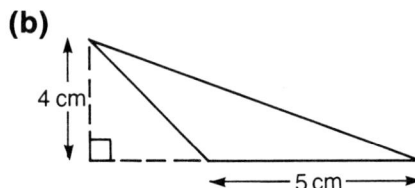

(a)

6 cm

3 cm

(b)

4 cm

5 cm

4. Copy these and fill in the gaps:

(a) 2000 ml = _____ l **(b)** 4.6 l = _____ ml

5. Work out these amounts:

(a) 5% of £2 **(b)** 30% of £10 **(c)** 20% of 50p

6. Ribbon costs 12p per yard. How much will it cost to buy:

(a) 1 ft? **(b)** 4 ft? **(c)** 6 ft?

7. Draw a grid on centimetre squared paper.
Mark it to 10 across and up.
Plot these points and join them up in order:

(1,9) (1,7) (4,7) (4,1) (6,1) (6,7) (9,7) (9,9) (1,9)

8. Tricia is making steamed pudding. She needs 3 teaspoons baking powder for each 8 oz plain flour.

How much should she use for **(a)** 4oz flour? **(b)** 16 oz flour?

9. Measure these angles:

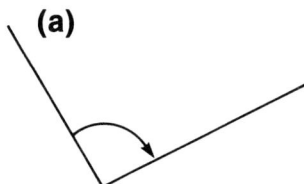

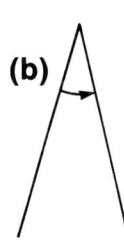

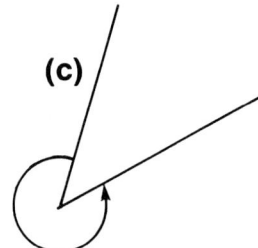

(a)   **(b)**   **(c)**

Percentages to fractions

'Per cent' means 'out of 100'.
We can write this as a fraction:

$25\% = \dfrac{25}{100}$ twenty-five out of 100

We can use equivalent fractions to cancel $\dfrac{25}{100}$ to its lowest terms.

$$\overset{\div 5 \quad \times 5}{\dfrac{25}{100} = \dfrac{5}{20} = \dfrac{1}{5}}_{\div 5 \quad \div 5}$$

$100\% = \dfrac{100}{100}$: 1 whole one

Exercise 1

Change these percentages to fractions.
The first one has been done for you.

$$\textbf{1.}\; 40\% = \overset{\div 10 \quad \times 2}{\dfrac{40}{100} = \dfrac{4}{10} = \dfrac{2}{5}}_{\div 10 \quad \div 2} \qquad 40\% = \dfrac{2}{5}$$

Now you do the rest.

2. 50%	**7.** 60%	**12.** 100%
3. 10%	**8.** 70%	**13.** 95%
4. 20%	**9.** 80%	**14.** 1%
5. 30%	**10.** 90%	**15.** 5%
6. 75%	**11.** 45%	

16. In the sale, a shop reduces everything by 25%.
  (a) What fraction of the price is taken off?
  (b) What percentage of the old price is the sale price?
  (c) What fraction of the old price is this?

17. Jane eats 50% of a loaf of bread.
  What fraction of the loaf does she eat?

18. Paulette spends 75% of her money on records.
  What fraction of her money is spent on records?
  What percentage of her money is spent on other things?

Perimeters

The **perimeter** of a shape is the distance all the way around it.

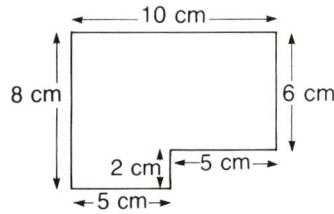

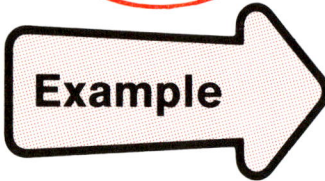

Perimeters of rectangles are easy:

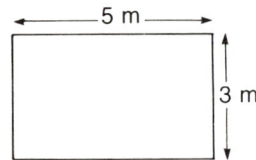

Opposite sides are the same length.
Each length counts twice.
So the perimeter is $2 \times 5 + 2 \times 3$

$$= 10 + 6 = 16 \text{ m}.$$

Squares are even easier:

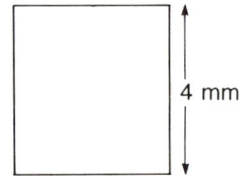

All the sides are the same length.
There are four sides.
So the perimeter is $4 \times 4 = 16$ mm.

Exercise 1

Work out the perimeters of these shapes:

1. 3 cm

2. 6 cm, 4 cm

3. 4 m, 1 m, 2 m, 2 m, 6 m

4. 30 m, 10 m

5. 8 mm, 6 cm, 4 mm, 4 mm, 10 mm, 4 mm

6. 20 cm

Sometimes you need to work out some of the lengths yourself.

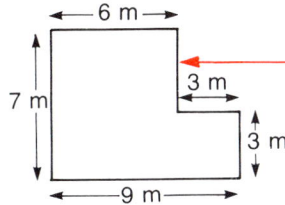

Example

We need to work out this length.

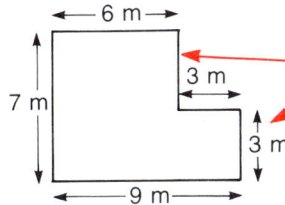

These two lengths must add up to 7 m.
So the missing length is 4 m.

The perimeter is 4 + 3 + 3 + 9 + 7 + 6 = 32 m.

Exercise 2

Find the missing lengths in these shapes and work out their perimeters. Some have more than one missing length.

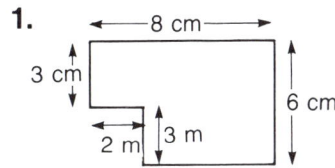

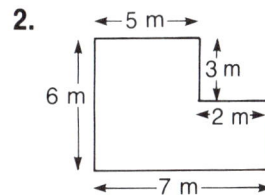

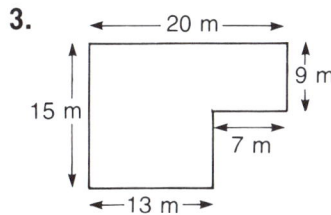

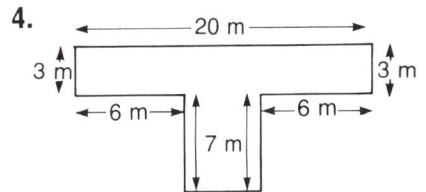

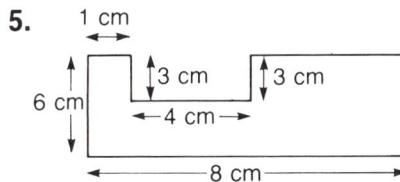

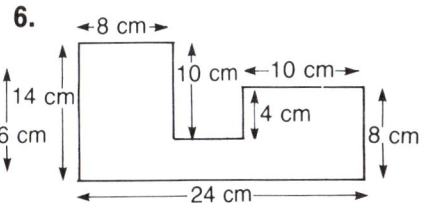

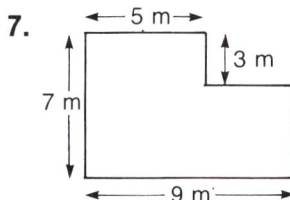

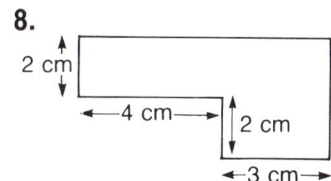

1.

2.

3.

4.

5.

6.

7.

8.

Changes

Alice is recording the weather for geography.
She records the temperature every day at midday for two weeks.

Here are her results:

day	1st week	2nd week
Sunday	18°C	33°C
Monday	21°C	30°C
Tuesday	20°C	26°C
Wednesday	23°C	28°C
Thursday	27°C	23°C
Friday	28°C	20°C
Saturday	30°C	22°C

Alice draws a line graph to display her results.
She uses squared paper to help her.

Exercise 1

Follow the steps below to draw Alice's graph yourself.

1. Draw two axes and label them.
Mark off every 2°C from 16° on the vertical axis.

2. Mark off the days on the horizontal axis.
Mark off on the *lines*.

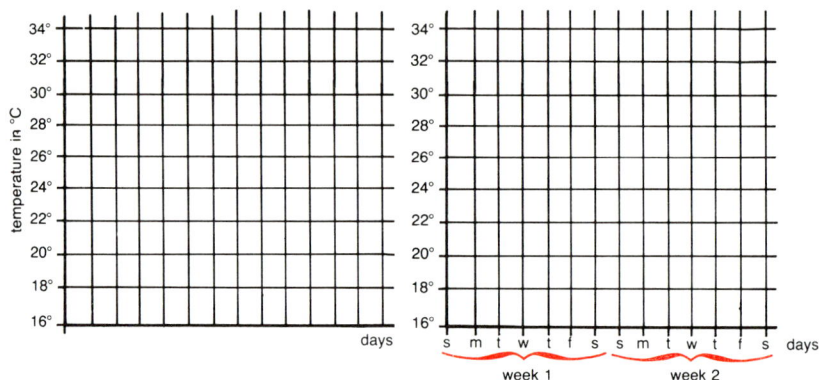

3. Plot the temperature each day with a cross. Use a sharp pencil.

4. Join up the points using a ruler. Give the graph a title.

Temperature changes over 2 weeks

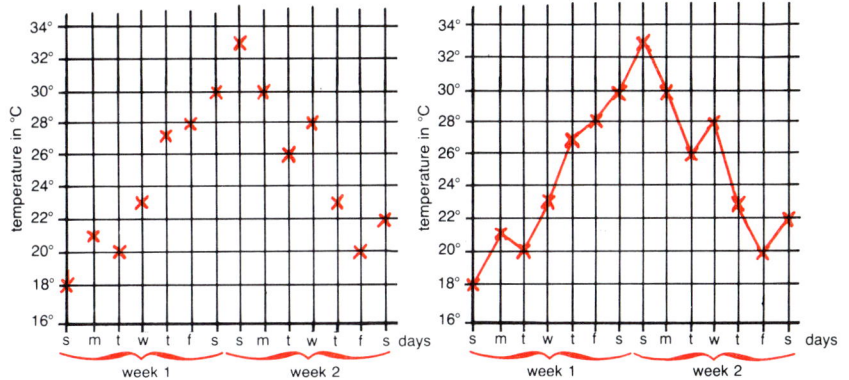

2. Use your graph to answer these questions:

(a) Which day was it hottest?
(b) Which day was it coolest?
(c) During the two weeks there were six days when it got hotter every day. Which days were they?

3. Here is a chart showing sales of Chocco bars over a year.

month	sales in millions
January	4.0
February	6.0
March	5.0
April	5.5
May	4.0
June	4.5
July	4.0
August	3.5
September	6.5
October	7.0
November	8.0
December	10.5

Draw a line graph to show this information.
Mark off every 1.0 million on the vertical axis.
Start at 3.0 million.

4. Here is a temperature chart for a hospital patient. Use it to answer the questions below.

Temperature Record: Jane Briggs

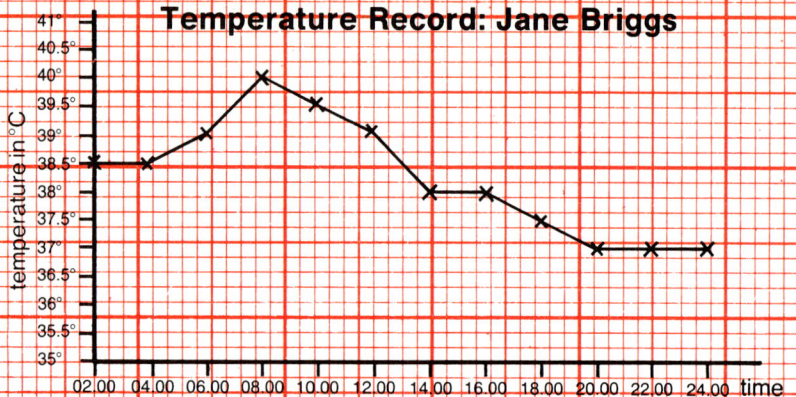

(a) How often is the patient's temperature taken?
(b) What is her temperature at 08.00?
(c) What is her temperature at 18.00?
(d) Normal body temperature is about 37°C. What time does the patient's temperature return to normal?
(e) When does the patient's temperature start to come down?

5. Here is a chart showing the weight of a newborn baby over eight weeks. She is weighed when she is born and then once a week.

date	Weight in kg
10 1.85	3.5
17 1.85	3.3
24 1.85	3.1
1 2.85	3.4
8 2.85	3.8
15 2.85	4.1
22 2.85	4.5
1 3.85	4.8

Note: It is normal for newborn babies to lose weight over the first 10 days after they are born!

Draw a line graph to show how the baby's weight changes. Mark off every 0.1 kg on the vertical axis. Start at 3.0 kg.
Mark off every week on the horizontal axis.

Decimals to fractions

We can use columns to write decimals as fractions.
The column headings tell us what sort of fraction we have.

Example

Write 0.3 as a fraction.

Put in the columns:

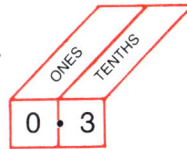

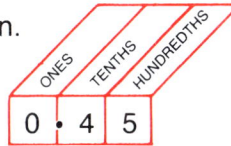

ONES	TENTHS
0 •	3

0.3 is 3 tenths. We write it $\dfrac{3}{10}$

Example

Write 0.45 as a fraction.

Put in the columns:

ONES	TENTHS	HUNDREDTHS
0 •	4	5

0.45 is 45 hundreds. We write it $\dfrac{45}{100}$

$\dfrac{45}{100}$ is not written in its lowest terms.

It must be cancelled.

$$\frac{45}{100} = \frac{9}{20} \quad (\div 5)$$

Exercise 1

Write these decimal numbers as fractions.
Cancel them to their lowest terms if you need to.

1. 0.5	**5.** 0.25	**9.** 0.05	**13.** 0.15
2. 0.2	**6.** 0.75	**10.** 0.01	**14.** 0.4
3. 0.6	**7.** 0.3	**11.** 0.8	**15.** 0.95
4. 0.9	**8.** 0.35	**12.** 0.85	**16.** 0.46

More flow diagrams

We can use flow diagrams to show what order to do things in.

Here is a flow diagram showing what you do when you play an LP.

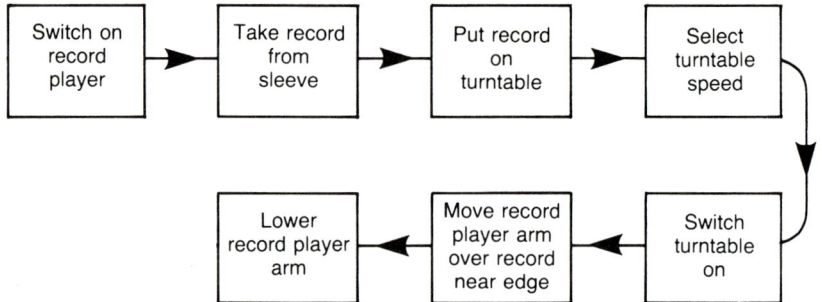

| Switch on record player | → | Take record from sleeve | → | Put record on turntable | → | Select turntable speed |

| Lower record player arm | ← | Move record player arm over record near edge | ← | Switch turntable on |

We can draw flow diagrams like this for many activities.
Some need more boxes than others.

Exercise 1

Draw flow diagrams to show how to do each of these things:

1. playing a cassette
2. throwing a ball
3. walking upstairs
4. making a cup of tea
5. putting on a shirt

6. What activity does this flow diagram describe?

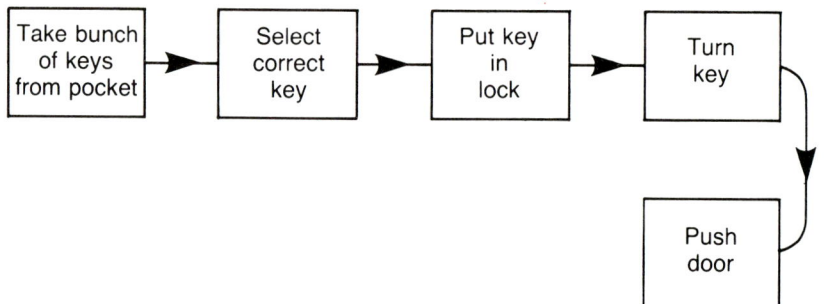

| Take bunch of keys from pocket | → | Select correct key | → | Put key in lock | → | Turn key |

| Push door |

Earlier in the book we worked out the answers to flow diagram sums.
In this section we are going to make the flow diagrams ourselves.

We use a letter instead of the starting number.

Example

$3a + 4$

This means: take a number
multiply by 3
add 4.

We make this flow diagram:

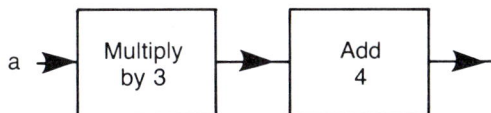

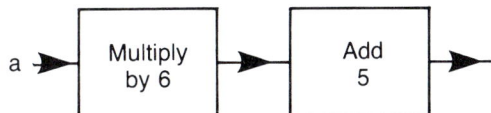

a → | Multiply by 3 | → | Add 4 | →

Example

We always start with the letter, even when it is at the end of the sum.

$5 + 6a$

This means: take a number
multiply by 6
add 5.

We make this flow diagram:

a → | Multiply by 6 | → | Add 5 | →

Exercise 2

Make flow diagrams for these sums:

1. $3a + 5$ **6.** $2a \times 5$

2. $2a - 3$ **7.** $4b - 3$

3. $5a \div 6$ **8.** $6b - 12$

4. $3 + 2a$ **9.** $7p \div 4$

5. $5 + 3b$ **10.** $3 + 5p$

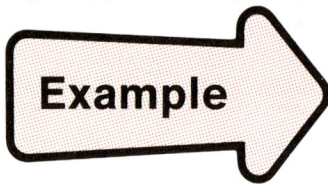

For some sums we need to use new phrases.

$3 - 2b$

This means: take a number
multiply by 2
take away from 3

We make this flow diagram:

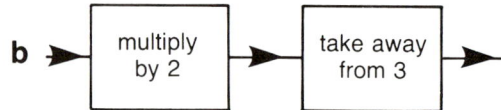

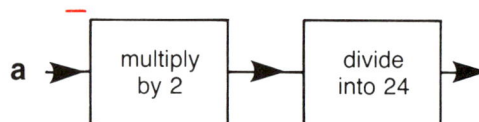

b → | multiply by 2 | → | take away from 3 | →

Exercise 3

Make flow diagrams for these sums.
Take care! Some use 'take away', some use, 'take away from'.

1. $15 - 3a$ **5.** $3b - 5$
2. $6 - 2a$ **6.** $2 - 4c$
3. $50 - 4b$ **7.** $6 - 2c$
4. $4b - 10$ **8.** $12c - 7$

We need a new phrase for division sums too.

$24 \div 2a$

This means: take a number
multiply by 2
divide into 24

We make this flow diagram:

a → | multiply by 2 | → | divide into 24 | →

Exercise 4

Make flow diagrams for these sums.
Take care! Some use 'divide by', some use 'divide into'.

1. $6 \div 2a$

2. $30 \div 3b$

3. $4b \div 6$

4. $28 \div 2c$

5. $16 \div 3a$

6. $54 \div 3b$

7. $16b \div 4$

8. $35 \div 6c$

Now make flow diagrams for these:

9. $3a + 5$

10. $4b - 2$

11. $6 - 3b$

12. $4a \div 6$

13. $3a \times 5$

14. $5 \times 2a$

15. $16 \div 2c$

16. $4b \div 3$

17. $9 - 3a$

18. $4a + 12$

19. $6 - 2c$

20. $5b \div 7$

When we use brackets we must do the sum inside the bracket first.

Example

$3 (2a + 4)$

This means: take a number
               multiply by 2
               add 4
               multiply the total by 3.

We make this flow diagram:

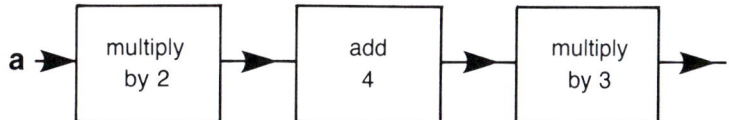

a → | multiply by 2 | → | add 4 | → | multiply by 3 | →

Exercise 5

Make flow diagrams for these sums.
Remember to do the sum inside the bracket first.

1. $2 (3a + 2)$

2. $3 (4p + 7)$

3. $(2a - 1) \div 3$

4. $(6 - a) \div 4$

5. $5 (4 - 3a)$

6. $27 \div (2a + 3)$

7. $40 \div (6 - 3c)$

8. $36 - (6 + 5b)$

9. $6 (56 - 4p)$

10. $50 - (2a + 7)$

Floor areas

Paul wants to carpet his bedroom floor.
He must work out how much carpet he needs.
Carpet is sold by the square metre.
Paul must work out the area of the floor in square metres.

Here is a diagram of the floor:

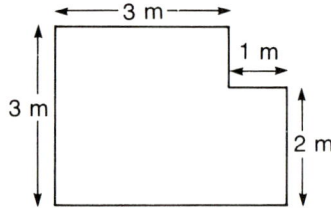

To work out the area, Paul must divide it into two parts:

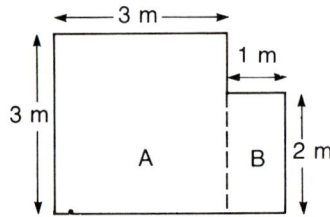

Paul's room is a square with
a rectangle on the side.
We can work out areas A and
B separately and add them
together.

Note: We could divide Paul's
bedroom like this:

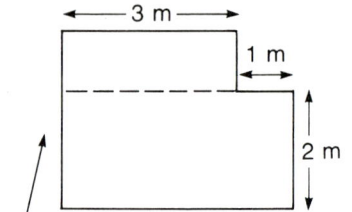

But we don't know the lengths
of all the sides.
The other way is easier.

Area of Paul's bedroom is Area A + Area B.

Area of a rectangle = length × width

Area A = $3 \times 3 = 9$ m^2
Area B = $2 \times 1 = 2$ m^2
So the total area = $9 + 2 = 11$ m^2.

Exercise 1

Work out the areas of these floors by dividing them into smaller parts. Make sure you divide them so you know the lengths of all the sides you need.
The first one has been done for you.

1.

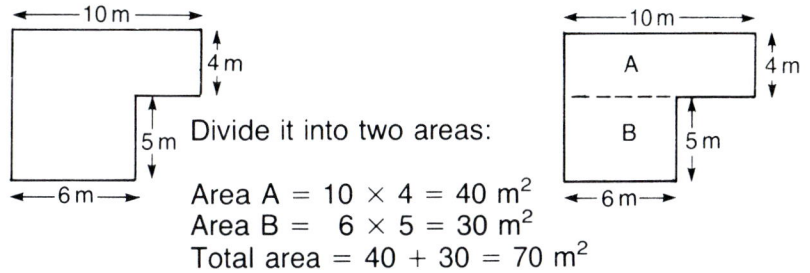

Divide it into two areas:

Area A = 10 × 4 = 40 m²
Area B = 6 × 5 = 30 m²
Total area = 40 + 30 = 70 m²

Now do the rest the same way.

2.

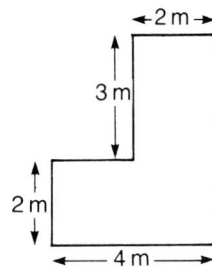

3.

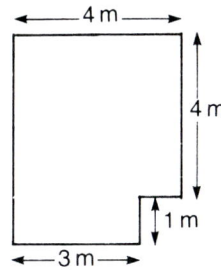

4.

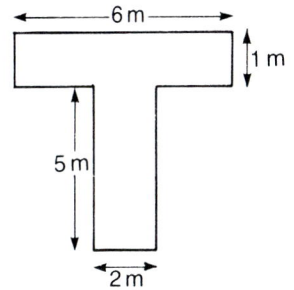

5.

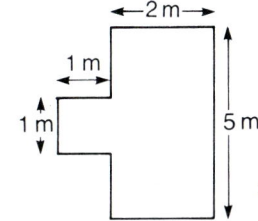

6.

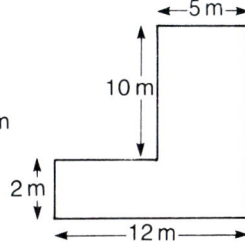

7.

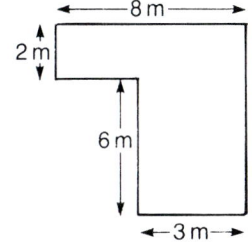

Some floors need to be divided into more than two parts. Try these ones.

8.

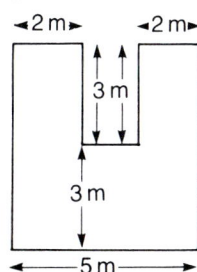

9.

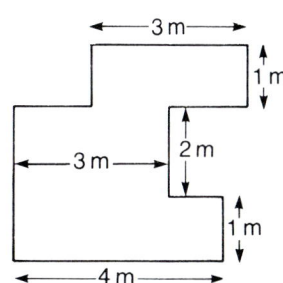

10.

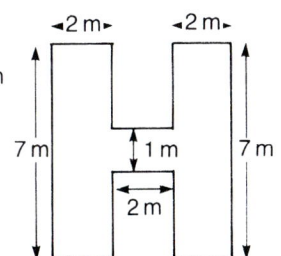

Averages

Roy plays cricket.
He wants to know how good a batsman he is compared to Barjinder.

They write down their scores for each innings they complete.

Roy's scores	Barjinder's scores
6	41
103	87
41	42
96	30
50	
64	

This does not tell them very much.
Roy has played more games than Barjinder.

They work out their batting averages to compare scores.

Batting average = $\dfrac{\text{total number of runs}}{\text{number of completed innings}}$

Altogether Roy scored 360
runs.
He had 6 innings.

So his batting average

$= \dfrac{360}{6} = 60$ runs

Altogether Barjinder scored 200
runs.
He had 4 innings.

So his batting average

$= \dfrac{200}{4} = 50$ runs

So Roy is a better batsman than Barjinder.

Exercise 1

Work out batting averages for these cricketers:

1. Sanjeev	2. Eddy	3. Maxine	4. Stephen	5. Gulsen
34	16	39	106	18
16	82	51	97	121
28	14	0	34	53
56	35	23	89	0
2	29	89	101	6
44	53	14	42	0
	25		0	46
	18		11	60

Which person is the best at batting?

The batting average is the score a player would have got if he had scored the same number of runs in each complete innings.

We use averages for other things too.

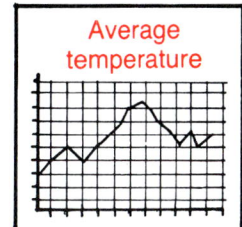

SWINE MATCHES

Average Contents: 200

Average temperature

DAILY BLAH

AVERAGE EARNINGS NOW £150

This sort of average is also called the **mean**

$$\text{mean} = \frac{\text{total}}{\text{number of items}}$$

Tricia wants to work out the mean amount she spends each week.

She records what she spends every week for 10 weeks.

Here are her results:

| £1 | £1.50 | £1.25 | £0.30 | £4.50 |
| £2.23 | £0.77 | £2.05 | £1.85 | £3.45 |

Altogether Tricia spends £18.90.

So the mean amount she spends per week is $\dfrac{£18.90}{10}$

= £1.89.

Exercise 2

Work out the mean for each set of numbers.

1. £1, £1.50, £2.50, £2
2. £3.50, £7.25, £2.25, £1.75, £7.50, £3.25
3. £2.60, £1.20, £4.10, £2.90, £6.20
4. £4.10, £0.20, £3.70, £0.90. £5.90, £3.20
5. 25p, 75p, 80p, 30p, 90p
6. 26p, 18p, 24p, 99p, 35p, 41p, 16p, 29p
7. 141, 236, 18, 410, 365, 612, 419, 207
8. 28, 31, 46, 34, 55, 18, 96, 21, 102, 39
9. 30 g, 15 g, 60 g, 95 g
10. 54 kg, 11 kg, 36 kg, 17 kg, 72 kg, 8 kg
11. Here are the scores of 10 pupils in a test:
    96, 84, 35, 41, 18, 67, 38, 86, 14, 72, 39
    What is the mean score?
12. Here are the ages of members of a family:
    39, 35, 16, 14, 12, 4
    What is the mean age?
13. Five people gave money to charity.
    Altogether they collected £3.75.
    What was the mean amount given?
14. Renu went cycling for 7 days.
    Altogether she travelled 315 miles.
    What was the mean distance travelled per day?

Pounds and ounces

We still use pounds (lb) and ounces (oz) for weighing many things.
Fruit and vegetables are sold by the pound.
Sweets and chocolates are often sold in quarter pound or half pound bags.

There are sixteen ounces in a pound.

$$16 \text{ oz} = 1 \text{ lb}$$
$$\text{So} \quad 8 \text{ oz} = \frac{1}{2} \text{ lb}$$
$$4 \text{ oz} = \frac{1}{4} \text{ lb}$$

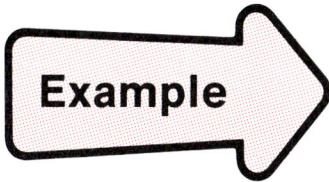

Often weights are written in pounds and ounces.
We can change them to ounces only.

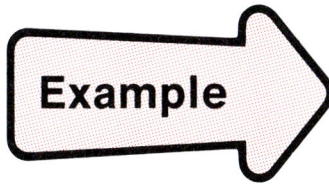

Example

2 lb 4 oz
2 lb = 2 × 16 =   32 oz
                  + 4 oz
                  ‾‾‾‾‾‾‾
                   36 oz

2 lb 4 oz = 36 oz

We can also change ounces to pounds and ounces.

Example

18 oz
18 oz = 1 × 16 + 2 oz left over
      = 1 lb 2 oz

18 oz = 1 lb 2 oz

Exercise 1

Now do these in the same way.

1. Change these to ounces only:

(a) 1 lb 3 oz **(b)** 1 lb 8 oz **(c)** 1 lb 6 oz **(d)** 2 lb
(e) 2 lb 4 oz **(f)** $\frac{3}{4}$ lb **(g)** 1 lb 14 oz **(h)** 2 lb 12 oz

2. Change these to pounds and ounces:

(a) 20 oz **(b)** 24 oz **(c)** 32 oz **(d)** 36 oz
(e) 30 oz **(f)** 40 oz **(g)** 28 oz **(h)** 23 oz

Drawing angles

We use a **protractor** to draw angles.
Here is a picture of one.

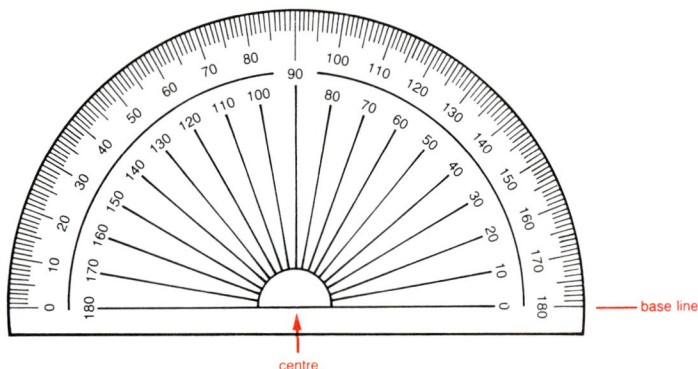

Notice: The base line is not quite at the bottom.
The protractor is marked in two directions from
0° to 180°.

Follow these steps to draw an angle.
We are going to draw an angle of 75°.

1. Draw a straight line.
Mark a point A near one end.

2. Put your protractor on the line.
Line up the base line with the line you drew.
Make sure the centre line is on the point you marked.

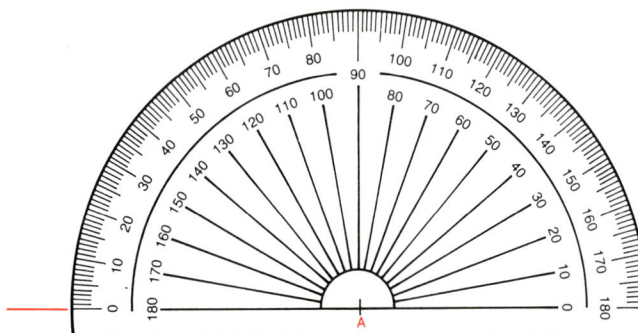

3. Count round the protractor from 0° to 75°.
Put a dot at 75°.

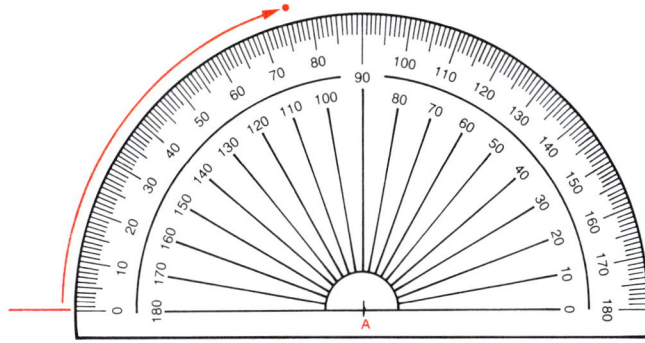

4. Take the protractor away.
Draw a straight line from A through the dot.
Mark which side of the line the angle is.

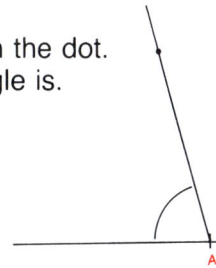

Exercise 1

Draw these angles in the same way:

1. 30°   **4.** 45°   **7.** 135°
2. 60°   **5.** 65°   **8.** 160°
3. 20°   **6.** 120°   **9.** 149°

Sometimes we need to draw an angle the other way round.
Follow these steps to draw an angle of 116°.

1. Draw a straight line.
Mark a point A near one end.

3. Line your protractor up as before.
Count round from 0° to 116°.
Put a dot at 116°.

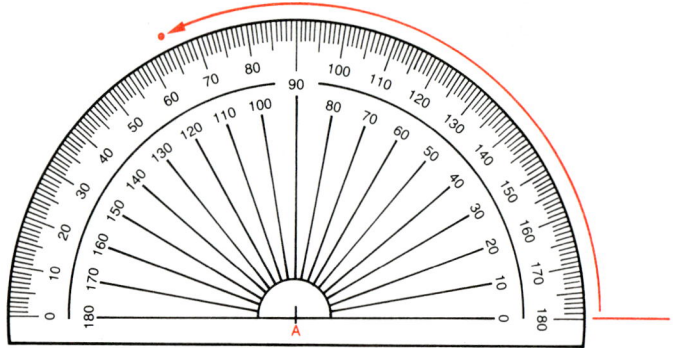

3. Take your protractor away.
Draw a straight line from A through the dot.
Mark which side of the line the angle is.

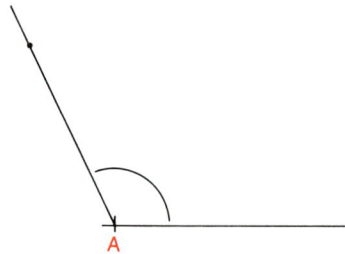

Exercise 2

Draw these angles in the same way:

1. 70°	**4.** 162°	**7.** 137°
2. 150°	**5.** 81°	**8.** 98°
3. 37°	**6.** 23°	**9.** 46°

Shovels

Maxine's brother Kevin works for a builder.
He often has to mix concrete.

Kevin is mixing 1 : 4 concrete for a wall.
This means that for every shovel of cement he uses 4 shovels of sand.

Kevin uses 2 shovels of cement.
So he should use $4 \times 2 = 8$ shovels of sand.

Exercise 1

A. How much sand will Kevin need for these amounts of cement?

1. 1 shovel   **4.** 4 shovels
2. 3 shovels   **5.** 5 shovels
3. 10 shovels   **6.** 8 shovels

B. For paving bricks Kevin has to mix 1 : 3 concrete.
For every shovel of cement he uses 3 shovels of sand.

How much sand will he need for these amounts of cement?

1. 1 shovel   **4.** 10 shovels
2. 2 shovels   **5.** 5 shovels
3. 4 shovels   **6.** 20 shovels

C. How much cement will Kevin need to make 1 : 3 concrete if he uses these amounts of sand?

1. 3 shovels   **4.** 18 shovels
2. 6 shovels   **5.** 9 shovels
3. 21 shovels   **6.** 24 shovels

Check-up 7

1. Work out the perimeters of these shapes:

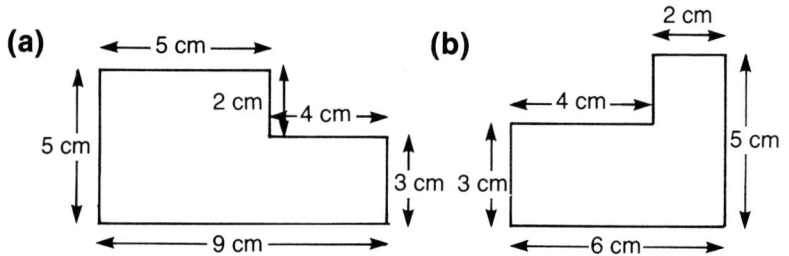

(a)

(b)

2. Here is a chart showing takings at the school tuck shop one week.
Draw a line graph to show this information.
Mark off every 50p on the vertical axis.
Start at £10.00.

day	takings
Monday	£10.80
Tuesday	£12.45
Wednesday	£16.30
Thursday	£11.40
Friday	£18.95

3. Change these percentages to fractions:
   (a) 20%   **(b)** 75%   **(c)** 7%

4. Change these decimals to fractions:
   (a) 0.6   **(b)** 0.25   **(c)** 0.08

5. Draw flow diagrams for these sums:
   (a) $3a - 4$ **(b)** $15 - 2a$ **(c)** $24 \div 3a$ **(d)** $2(2a \div 3)$

6. Work out the areas of these floors:
   (a)

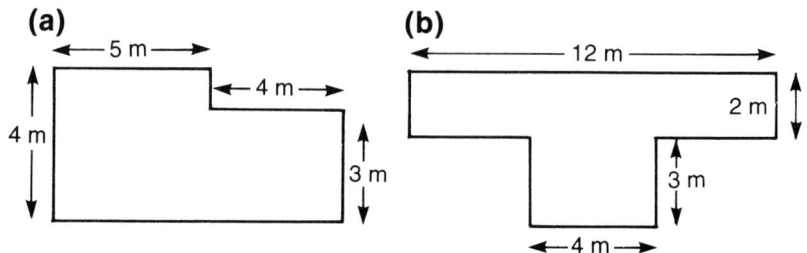

   (b)

7. Work out the mean of these numbers: 14, 17, 12, 9, 28.

8. Kevin is mixing 1 : 4 concrete. For every shovel of cement he uses 4 shovels of sand.
   How much cement will he use for these amounts of sand?
   (a) 8 shovels   **(b)** 12 shovels

9. Draw these angles:
   (a) 85°   **(b)** 120°   **(c)** 12°

Solids

Shapes like squares and triangles are flat on the page.
Solid shapes can be picked up and thrown around.

We can find out more about solids by making some.

Exercise 1

The shapes below and on the next page are nets of solids.
They can be folded and stuck to make solid shapes.

Trace the nets onto stiff paper or thin card.
Do it as accurately as you can. Use a ruler.
Cut them out along the solid lines: _____
Fold them along the dotted lines: _ _ _ _ _ _ _
Stick the flaps under the edges with the same letters.
Write the name of each solid on it.

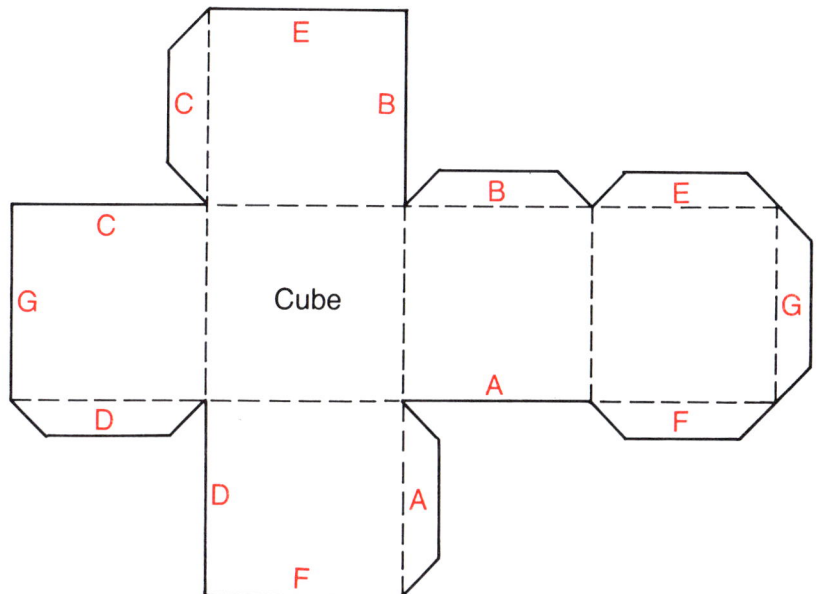

Cube

Cuboid

G

B

A

A

B

E

F

Cuboid

D

C

C

D

E

F

G

Tetrahedron

B

A

B

A

C

A

C

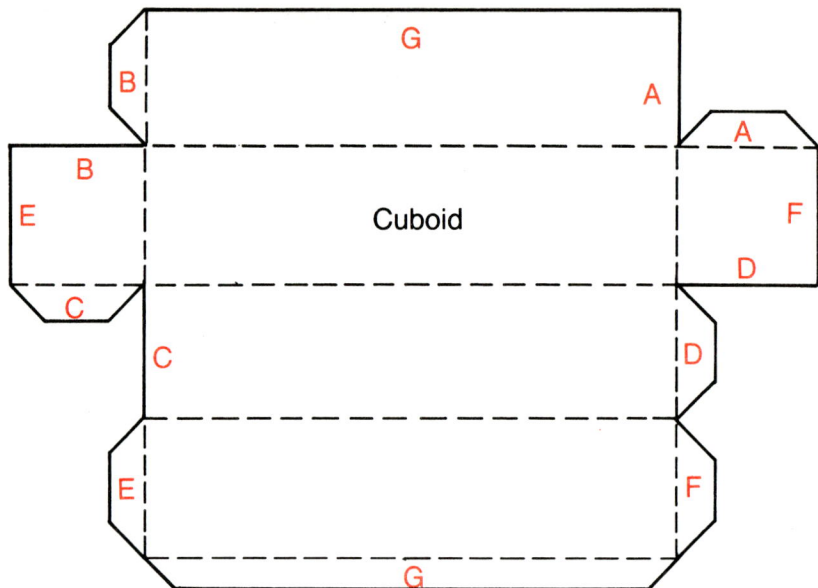

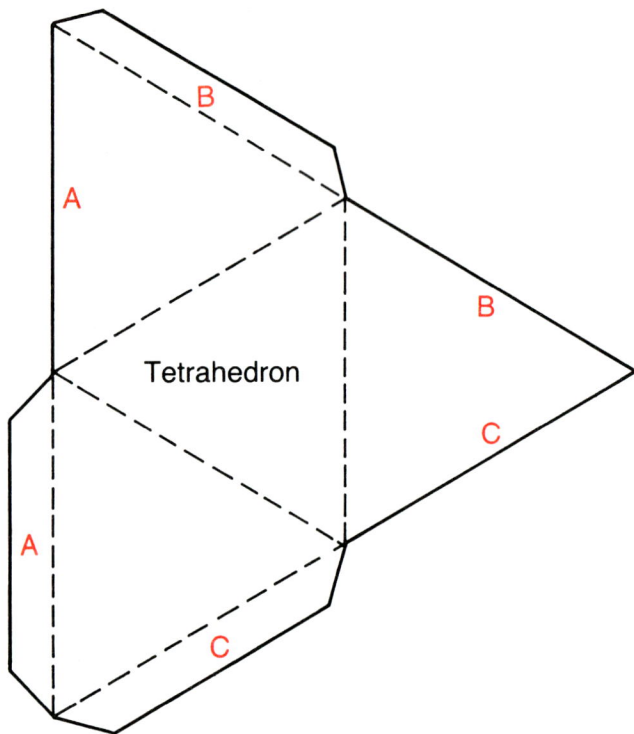

We cannot make a model of a **sphere**.
But there are lots of them all around us.
A rubber ball is a sphere.
So is a marble.

We also have to be able to name drawings of solids.

Here is a drawing of a **cube**.
The parts of it have been labelled.

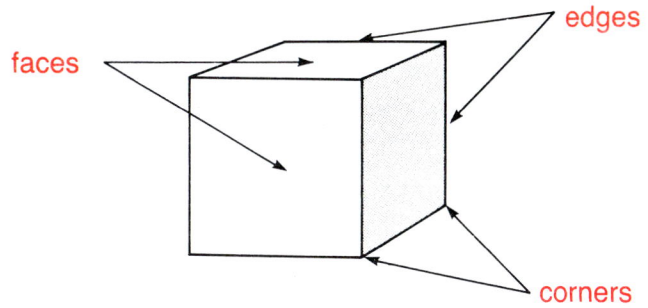

Exercise 2

Copy the table below into your book.
Use your solid shapes to help you complete it.

drawing	name	shape of face	number of faces	number of sides	number of corners
			6		
				12	
		equilateral triangle			

Rounding

4G have medicals at school.
The nurse weighs them.

Roy weighs 58.4 kg.
The nurse writes down 58 kg.
She has rounded Roy's weight to the nearest kilogram.
58.4 is nearer to 58 than 59.
So Roy is 58 kg to the nearest whole kilogram.

Darren weighs 50.8 kg.
50.8 is nearer to 51 than 50
So Darren weighs 51 kg to the nearest whole kilogram.
The nurse writes down 51 kg.

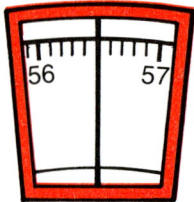

Avril weighs 56.5 kg.
56.5 is halfway between 56 and 57
We say that 56.5 kg is 57 to the nearest whole kilogram.
The nurse writes down 57 kg.

Exercise 1

Here are the weights of some more members of 4G
Write them to the nearest whole kilogram.

Tricia	54.8 kg	Gulsen	52.9 kg
Barjinder	56.5 kg	Curtis	51.8 kg
Alice	52.1 kg	Gulduren	59.5 kg
Renu	53.4 kg	Maxine	57.3 kg
Eddy	51.6 kg	Sharon	54.5 kg
Brian	57.2 kg		

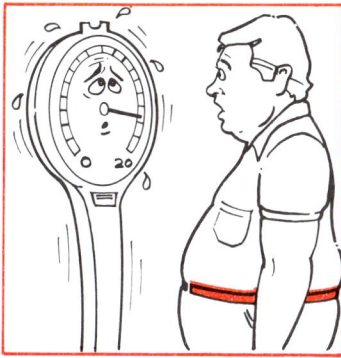

Numbers are often rounded off like this to help us see what size they are.

We can round off any decimal number to the nearest whole one.

We look at the tenths.
If the number is less than five, leave it off.
If it is five or more, round up to the nearest whole number.

Example

1. 13.2
      this is less than 5

   So 13.2 is 13 to the nearest whole one.

2. 293.5
      this is 5 or more

   So 293.5 is 294 to the nearest whole one.

Exercise 2

1. The nurse also measured 4G.
   Round their heights off to the nearest centimetre.

Roy	161.7 cm	Eddy	159.5 cm
Darren	159.4 cm	Brian	161.6 cm
Avril	160.5 cm	Gulsen	160.8 cm
Tricia	159.8 cm	Curtis	159.9 cm
Barjinder	160.9 cm	Gulduren	162.2 cm
Alice	159.3 cm	Maxine	160.7 cm
Renu	160.2 cm	Sharon	158.3 cm

2. Round these numbers to the nearest whole one:

117.2	215.3	316.2
36.8	420.5	1001.6
14.4	554.8	253.9

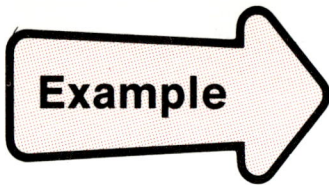

We often round numbers to the nearest 10.

1. 203

203 is nearer to 200 than 210.

So 203 = 200 to the nearest 10.

2. 167

167 is nearer to 170 than 160.

So 167 = 170 to the nearest 10.

Exercise 3

Round these numbers to the nearest 10:

36	146	375	1072
48	127	416	1346
72	103	589	2135
18	112	411	4717

We can also round numbers to the nearest 100.

1. 342

342 is nearer to 300 than 400.

So 342 = 300 to the nearest 100.

2. 756

756 is nearer to 800 than 700.

So 756 = 800 to the nearest 100.

Exercise 4

Round these numbers to the nearest 100:

218	576	763	729
499	342	1236	853
304	450	4890	249
421	835	5621	1793

Gallons and pints

We still use gallons (gal) and pints (pt) for measuring some liquids.

Petrol is sold by the gallon.
Milk and beer are sold by the pint.

There are eight pints in a gallon.
8 pt = 1 gal

So 2 gallons = 2 × 8 = 16 pt
   3 gallons = 3 × 8 = 24 pt

Exercise 1

Copy these and fill in the gaps:

1. 16 pt = _____ gal

2. 24 pt = _____ gal

3. 4 pt = _____ gal

4. 2 pt = _____ gal

5. 12 pt = _____ gal _____ pt

6. 18 pt = _____ gal _____ pt

7. 20 pt = _____ gal _____ pt

8. 10 pt = _____ gal _____ pt

9. 2 gal = _____ pt

10. 4 gal = _____ pt

11. 10 gal = _____ pt

12. 3 gal 2 pt = _____ pt

13. 2 gal 4 pt = _____ pt

14. 1½ gal = _____ pt

15. 1¼ gal = _____ pt

Missing angles

A straight line is one half turn.
So a straight line = 180°.

Angles on a straight line always
add up to 180°.

We can use this fact to find missing angles.

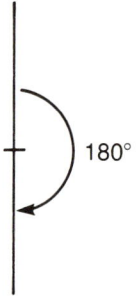
180°

Example

$70° + A = 180°$

So $A = 180° - 70°$

$A = 110°$

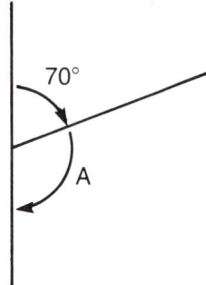

Example

$40° + 55° + B = 180°$

$\quad 95° + B = 180°$

So $B = 180° - 95°$

$B = 85°$

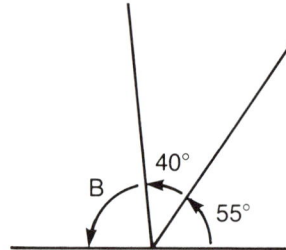

Exercise 1

Find these missing angles in the same way.

1.

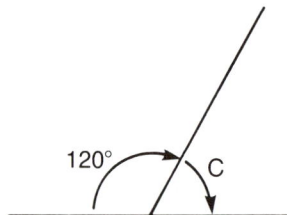

2.

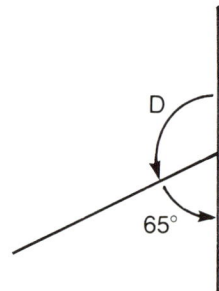

3.

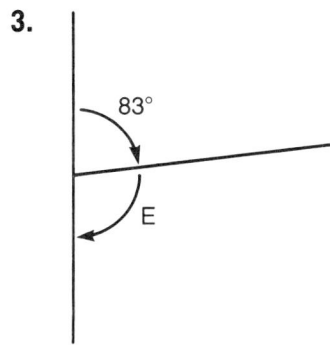

83°

E

4.

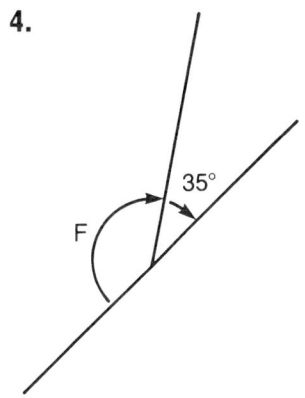

35°

F

5.

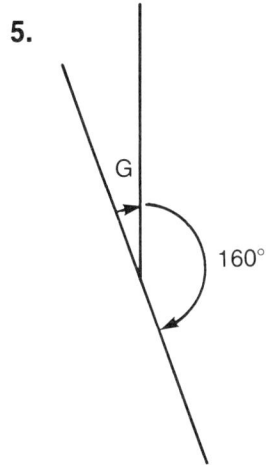

G

160°

6.

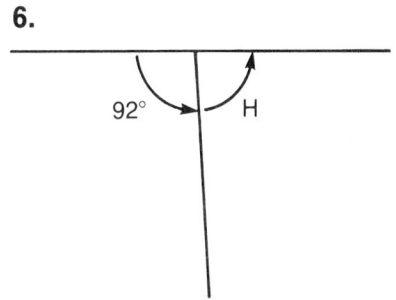

92°   H

7.

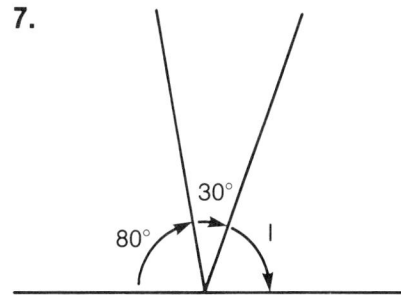

30°

80°   I

8.

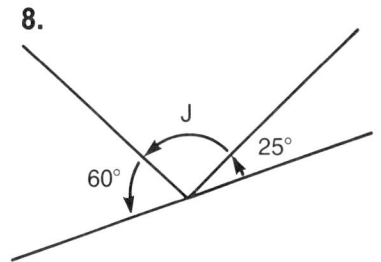

J

25°

60°

9.

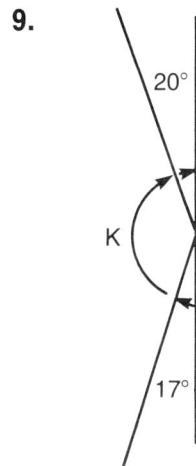

20°

K

17°

10.

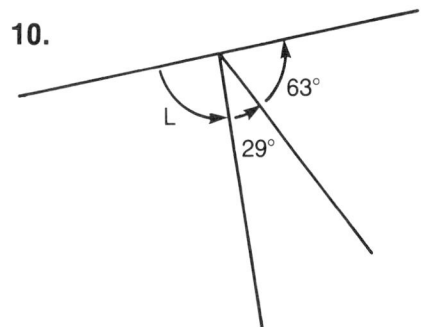

63°

L

29°

One full turn is 360°.

So angles in a full turn add up to 360°.

We can use this fact to find other missing angles.

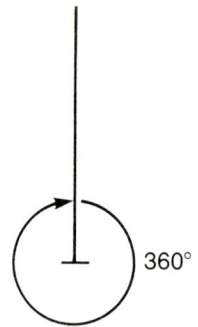
360°

$120° + A = 360°$

So $A = 360° - 120°$

$A = 240°$

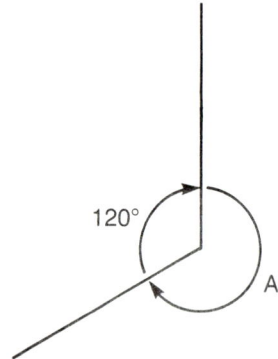
120°
A

$80° + 112° + B = 360°$

$192° + B = 360°$

So $B = 360° - 192°$

$B = 168°$

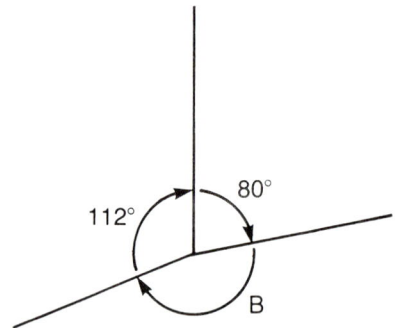
112°
80°
B

Find these missing angles in the same way.

1.

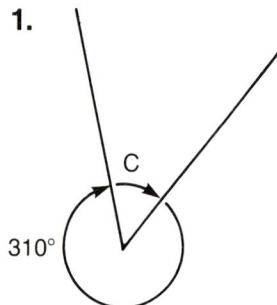
C
310°

2.

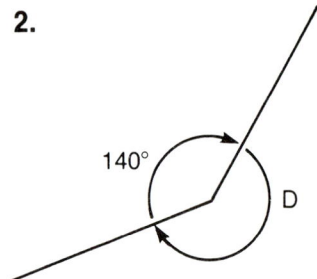
140°
D

166

3.

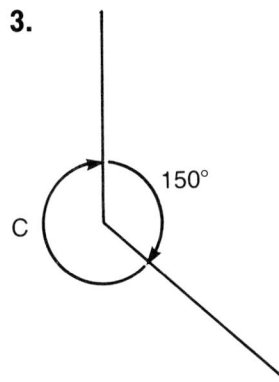

150°

C

4.

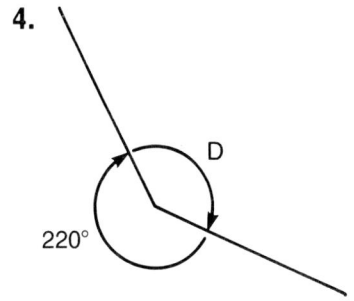

D

220°

5.

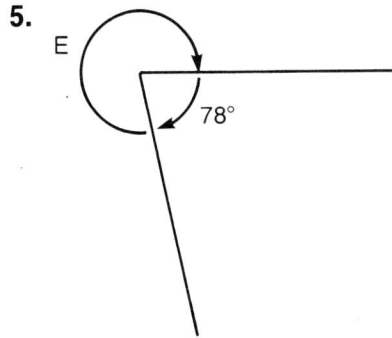

E

78°

6.

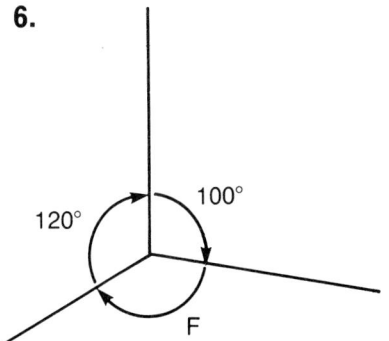

120° 100°

F

7.

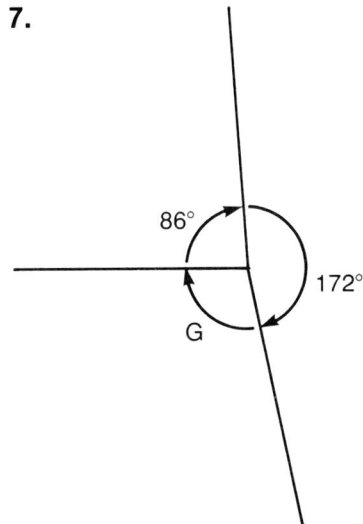

86°

172°

G

8.

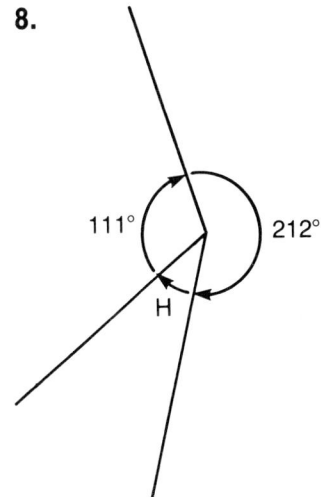

111° 212°

H

9.

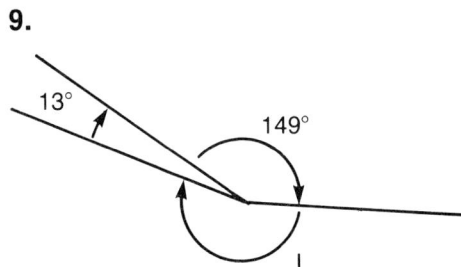

13° 149°

I

Flow diagrams and equations

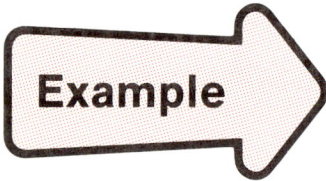

We can use flow diagrams to solve equations.
We make a flow diagram which undoes the equation.

$3a + 4 = 10$

We make this flow diagram:

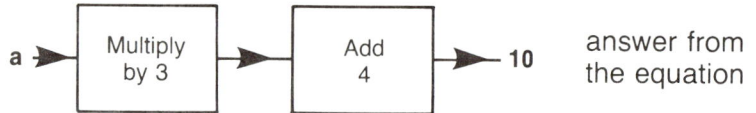

a → | Multiply by 3 | → | Add 4 | → 10   answer from the equation

We make a flow diagram to undo the equation.
Each box contains the opposite of what is in the original flow diagram.

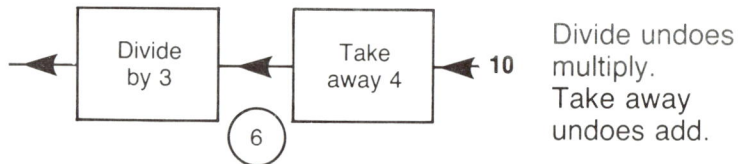

← | Divide by 3 | ← | Take away 4 | ← 10   Divide undoes multiply.
                (6)   Take away undoes add.

We do the sums in the undoing flow diagram to find the value of **a**.
We use the answer from the equation as the starting number.

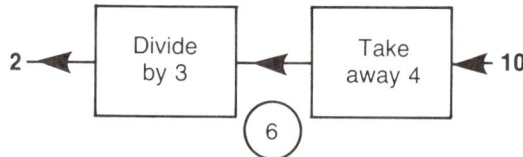

2 ← | Divide by 3 | ← | Take away 4 | ← 10
                  (6)

The answer is 2, so a = 2.

$\dfrac{a}{2} - 3 = 2$

flow diagram

a → | Divide by 2 | → | Take away 3 | → 2   Multiply undoes divide by.

10 ← | Multiply by 2 | ← | Add 3 | ← 2   Add undoes take away.
                    (5)

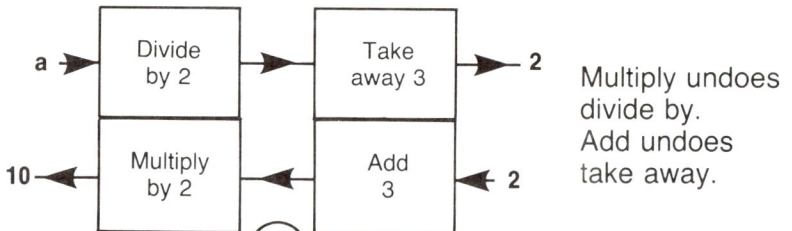

undoing flow diagram

The answer is 10, so a = 10.

Exercise 1

Use flow diagrams to solve these equations.
The first one has been done for you.

1. 4p + 1 = 13

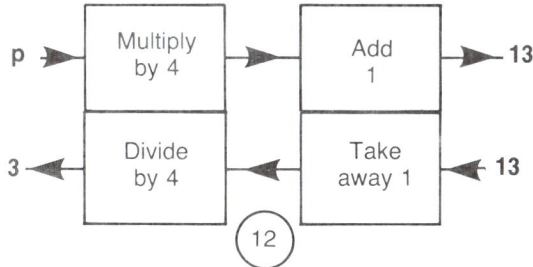

So p = 3.

Now you do the rest the same way.
Some need more than two boxes in each row.

2. 3r + 4 = 16
3. 2p − 5 = 11
4. 6r − 4 = 32
5. $\dfrac{r}{3}$ = 9
6. 4p + 7 = 27
7. $\dfrac{m}{6}$ + 4 = 10
8. $\dfrac{m}{4}$ − 8 = 32

9. 7x − 16 = 33
10. 6a + 50 = 92
11. $\dfrac{3p}{2}$ = 3
12. 5r − 16 = 4
13. $\dfrac{2m}{3}$ + 4 = 8
14. 6r − 4 = 20
15. $\dfrac{3x}{4}$ − 3 = 12

Example

We can check the answers by using them as the starting numbers in the original flow diagram.

In question 1, we got the answer p = 3.

We can check this answer in the flow diagram:

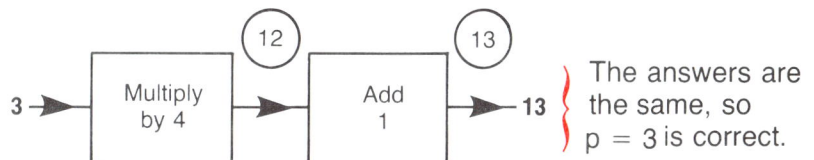

The answers are the same, so p = 3 is correct.

Now check the rest of your answers to Exercise 1 in the same way.

Take away from and divide into do not have opposites.
They stay the same in the undoing flow diagram.

$24 - 3a = 12$

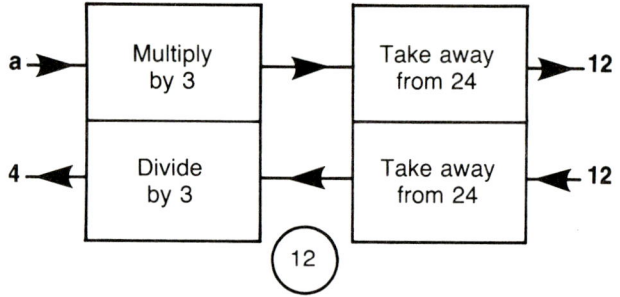

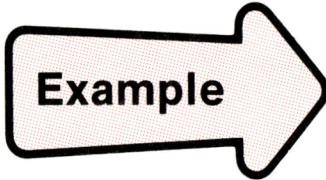 **Example**

So $a = 4$.

$36 \div 3p = 6$

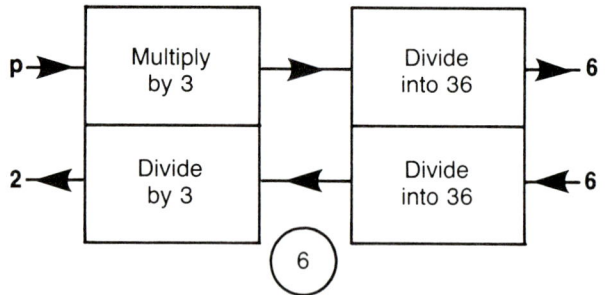

So $p = 2$.

Exercise 2

Use flow diagrams to solve these equations.

1. $4 - 2a = 2$

2. $30 - 5p = 15$

3. $16 \div 2p = 2$

4. $40 \div 4a = 10$

5. $36 \div 3r = 4$

6. $16 - 6s = 10$

7. $25 - 4x = 1$

8. $43 - 3y = 34$

9. $18 \div 2y = 3$

10. $25 - 4x = 9$

Summary

We can solve equations by using flow diagrams to undo them.

Take away undoes add
Add undoes take away
Divide undoes multiply by
Multiply by undoes divide by
Divide into undoes divide into
Take away from undoes take away from

Exercise 3

Use flow diagrams to solve these equations.

Remember: when there is a bracket, do the sum inside the bracket first.

1. $3a - 6 = 12$

2. $4p + 16 = 32$

3. $5a \div 4 = 5$

4. $3p - 21 = 15$

5. $54 - 2p = 48$

6. $3(5 + 2a) = 27$

7. $5(4p - 3) = 25$

8. $2(6 - p) = 10$

9. $3(15 - 3p) = 36$

10. $(2 + 3a) \div 7 = 2$

11. $(50 - 2a) \div 2 = 15$

12. $7(6 - 2x) = 28$

13. $4x - 17 = 31$

14. $27 - 3x = 15$

15. $2(10 - 2y) = 8$

16. $35 \div p = 5$

17. $40 \div 2p = 5$

18. $39 \div 3p = 13$

19. $4(12 \div p) = 12$

20. $2(18 \div 3p) = 6$

You've bitten off more than you can chew

Angles of a triangle

Draw a triangle on a piece of paper.
Cut it out.

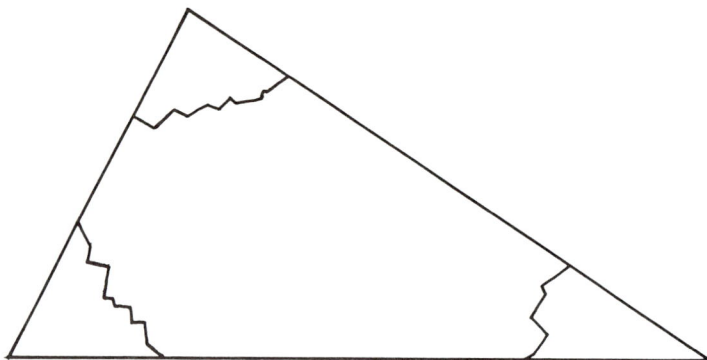

Tear off the corners.
See if you can arrange them along a straight line like this:

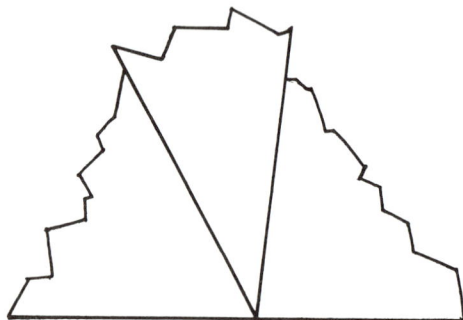

Do the same with two more triangles.
Look at your and your friends' triangles.
All the triangles are different.
All the corners can be arranged along a straight line.

The angles of any triangle can be put together to make a straight line.

A straight line is one half turn.
One half turn is 180°

So the angles of any triangle add up to 180°.

We can use this fact to find the third angle of a triangle if we know the other two angles.

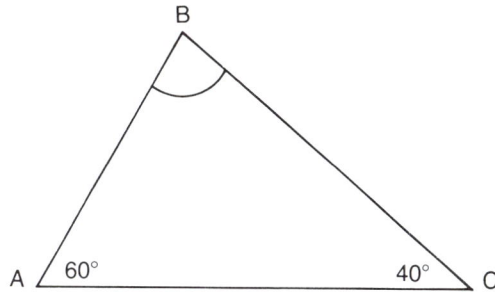

Example

We want to find angle B.

The angles all add up to 180°.
Angle A + Angle C = 60° + 40° = 100°

Angle B = 180° − 100° = 80°

Exercise 1

Find the third angle in these triangles:

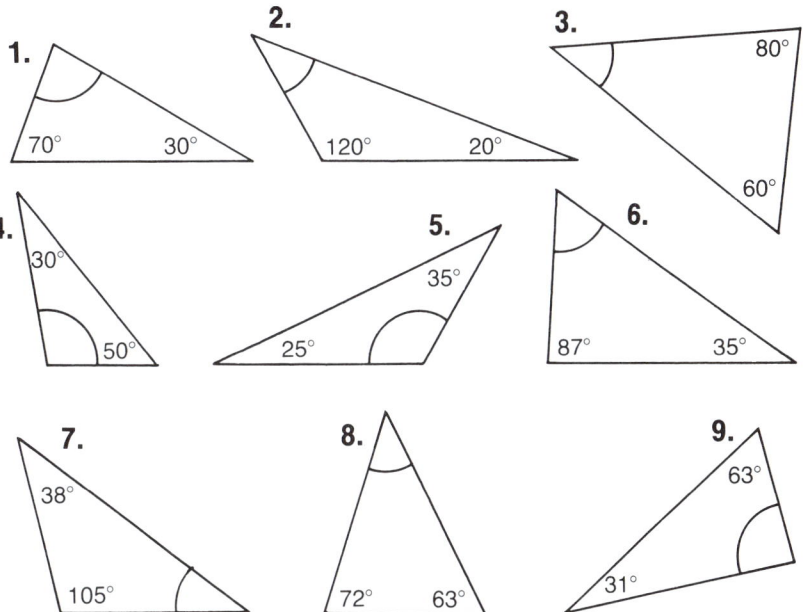

1.

70°   30°

2.

120°   20°

3.

80°

60°

4.

30°

50°

5.

35°

25°

6.

87°   35°

7.

38°

105°

8.

72°   63°

9.

63°

31°

173

Squares

Here is a square.

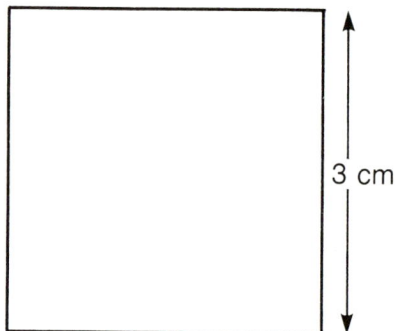

```
┌──────────────────────┐ ▲
│                      │ │
│                      │ │
│                      │ │
│                      │ │  3 cm
│                      │ │
│                      │ │
│                      │ │
└──────────────────────┘ ▼
```

Each side is 3 cm long.
The area of the square is 3×3 cm^2
$= 9$ cm^2

We call 3×3 3 squared

3 squared is written 3^2
$$3^2 = 3 \times 3 = 9$$

Exercise 1

A. Work out these values.
The first one has been done for you.

1. 4^2
$4^2 = 4 \times 4 = 16$

Now you do the rest the same way.

2. 5^2	**5.** 10^2	**8.** 7^2
3. 2^2	**6.** 9 squared	**9.** 1^2
4. 6 squared	**7.** 12^2	**10.** 8 squared

B. Find the areas of squares with sides these lengths.
The first one has been done for you.

1. 4 cm
Area $= 4^2 = 4 \times 4 = 16$ cm^2

Now you do the rest the same way.

2. 2 cm	**5.** 5 mm	**8.** 11 cm
3. 3 m	**6.** 8 mm	**9.** 13 cm
4. 6 m	**7.** 10 m	**10.** 12 cm

Here is another square.

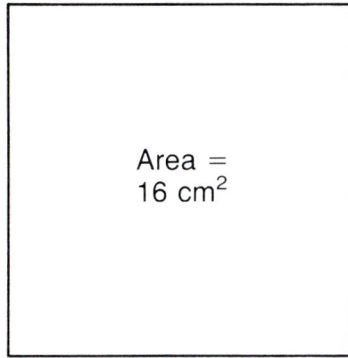

Area =
16 cm^2

The area is 16 cm^2.
How long are the sides?

Each side is 4 cm long because 4 × 4 = 16.

We say 4 is the square root of 16.
We write it $\sqrt{16}$.

16 means 'the number which gives you 16 when you multiply it by itself'.

Exercise 2

A. Here are the areas of some squares.
Find the lengths of their sides.
The first one has been done for you.

1. Area = 9m^2
Length of side = $\sqrt{9}$ = 3 m because 3 × 3 = 9

Now you do the rest the same way.

2. Area = 4 m^2 **5.** Area = 36 cm^2
3. Area = 49 m^2 **6.** Area = 81 cm^2
4. Area = 25 cm^2 **7.** Area = 144 m^2

B. Now work these out.

1. $\sqrt{9}$ **4.** $\sqrt{64}$
2. $\sqrt{1}$ **5.** $\sqrt{169}$
3. $\sqrt{100}$ **6.** $\sqrt{121}$

All mixed up

Whole numbers plus fractions, like $2\frac{3}{4}$, are called **mixed numbers**.

When we use them in calculations, it is easier if we write them just as fractions.

Reminder: $2\frac{3}{4} = \frac{8}{4} + \frac{3}{4} = \frac{11}{4}$ eleven quarters

When we have done this, we do the calculation in the usual way.

Example

$2\frac{2}{3} - 1\frac{1}{2}$

$2\frac{2}{3} = \frac{6}{3} + \frac{2}{3} = \frac{8}{3}$ $\qquad$ $1\frac{1}{2} = \frac{2}{2} + \frac{1}{2} = \frac{3}{2}$

We have $\frac{8}{3} - \frac{3}{2}$

We use equivalent fractions to convert $\frac{8}{3}$ and $\frac{3}{2}$ to sixths.

$\frac{8}{3} = \frac{16}{6}$ (×2) $\qquad$ $\frac{3}{2} = \frac{9}{6}$ (×3)

So we have $\frac{16}{6} - \frac{9}{6} = \frac{7}{6}$ which is $1\frac{1}{6}$

Exercise 1

Now try these.
If necessary, cancel your answers and write them as mixed numbers.

1. $1\frac{1}{2} + 2\frac{1}{4}$ $\qquad$ 5. $1\frac{1}{2} \times 2\frac{1}{4}$ $\qquad$ 9. $1\frac{3}{4} + 2\frac{1}{8}$

2. $2\frac{2}{3} - 1\frac{1}{4}$ $\qquad$ 6. $2\frac{1}{5} \times \frac{9}{10}$ $\qquad$ 10. $2\frac{1}{2} - 1\frac{3}{5}$

3. $1\frac{1}{8} + 2\frac{1}{2}$ $\qquad$ 7. $3\frac{1}{2} \div 1\frac{1}{4}$ $\qquad$ 11. $1\frac{1}{4} \times 2\frac{1}{2}$

4. $1\frac{2}{3} - \frac{4}{5}$ $\qquad$ 8. $1\frac{2}{3} \div 1\frac{5}{6}$ $\qquad$ 12. $2\frac{1}{3} \div 1\frac{1}{2}$

176

Check-up 8

1. Round these numbers to the nearest whole one:
   (a) 14.6   **(b)** 27.3   **(c)** 102.5

2. Round these numbers (a) to the nearest 10 (b) to the nearest 100:
   (a) 156   **(b)** 217   **(c)** 483

3. Copy these and fill in the gaps:
   (a) 8 pt = _____gal   **(b)** 2 gal 3 pt = _____pt

4. Find the missing angles:

   (a)   **(b)**

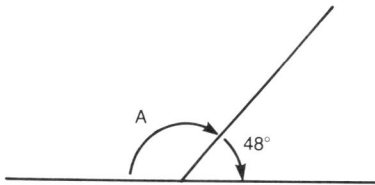
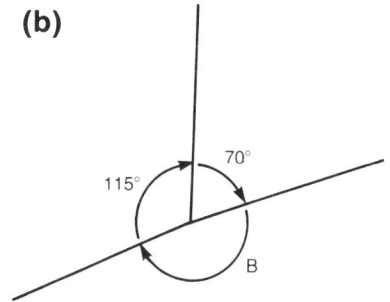

5. Use flow diagrams to solve these equations:
   (a) 3r + 5 = 14   **(b)** 10 − 2p = 2

6. Find the third angle in these triangles:

   (a)   **(b)**

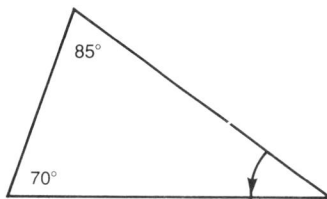
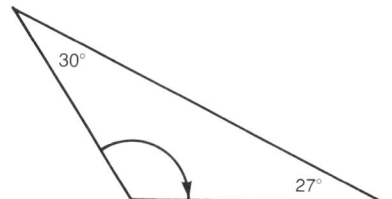

7. Work out these values:
   (a) 8^2 **(b)** 6^2 **(c)** 4 squared

8. Here are the areas of some squares.
   How long are the sides?
   (a) 4 m^2 **(b)** 36 m^2 **(c)** 25 cm^2

9. Work these out:

   (a) $2\frac{1}{2} + 3\frac{1}{4}$ **(b)** $4\frac{2}{3} - 2\frac{1}{6}$ **(c)** $1\frac{3}{5} \times 2\frac{7}{10}$

Calculator crossword

Work out the sums on your calculator, then turn the calculator upside down to read the answer to each clue.

Clues Across

1. $7.61 \div 100$
3. $188\ 303 \times 2$
7. 154×4
8. $6 \times 63\ 134$
11. $7053 + 665$
12. $1154 + 2222$
13. 257×3
15. $1 \div 2$
16. $1276 \div 2$
17. 1289×6
19. $15\ 469 \times 5$
22. 183.25×4
24. $158\ 000 + 158\ 008$
25. 2128.6×25
28. $100 - 62$
29. 809×7
31. 0.4×0.2
33. $7 \div 10$
35. $11\ 769 \times 32$
38. 2×0.25
39. 601×75
40. $2214 + 3393$
41. $10\ 000 - 4463$
42. $77.34 \div 100$
43. $556 + 51$

Clues Down

1. 3×257
2. 77×40
3. 368×2
4. 0.3×2
5. 141×27
6. 551×14
7. $6176 + 2000$
9. 6303×6
10. $5313 + 350$
14. 91×7
15. $106\ 090 \div 2$
18. 113×49
20. 12×67
21. $12\ 345 + 22\ 662$
23. $1614 \div 2$
26. $1 \div 50$
27. $6000 - 193$
28. 159.5×4
30. 81×75
32. 537.6×100
34. $6654 + 1084$
36. 5×8
37. 30×0.02
38. 0.141×5

Answers

Check your calculator work (page 5)

A.

1. 84	**3.** 495	**5.** 489	**7.** 740
2. 151	**4.** 253	**6.** 660	**8.** 1201

B.

1. 256	**3.** 103	**5.** 503	**7.** 104
2. 189	**4.** 228	**6.** 335	**8.** 61

C.

1. 912	**3.** 270	**5.** 1554	**7.** 1161
2. 432	**4.** 1272	**6.** 1980	**8.** 5824

D.

1. 45	**3.** 25	**5.** 15.5	**7.** 24
2. 48	**4.** 14	**6.** 36	**8.** 42

E.

1. 7.4	**3.** 27.1	**5.** 7.8	**7.** 3
2. 23.5	**4.** 11.6	**6.** 10.08	**8.** 29.6

Sharing it out (pages 6-8)
Exercise 1

B.

2. $\frac{1}{8}$	**4.** $\frac{1}{6}$	**6.** $\frac{1}{9}$	**8.** $\frac{1}{5}$
3. $\frac{1}{6}$	**5.** $\frac{1}{7}$	**7.** $\frac{1}{10}$	**9.** $\frac{1}{3}$

Exercise 2
A.

2. $\frac{2}{3}$	**3.** $\frac{4}{5}$	**4.** $\frac{3}{6}$	**5.** $\frac{1}{3}$

Numbers in columns (pages 9-11)
Exercise 1

A.

1. 600	**6.** 3 000 000	**11.** 500 046	**16.** 2001
2. 96	**7.** 13 000	**11.** 118	**17.** 50 080
3. 4000	**8.** 200 000	**13.** 7500	**18.** 50 008
4. 50 000	**9.** 150 000	**14.** 900 090	**19.** 3 700 000
5. 5 000 000	**10.** 9 000 000	**15.** 6 005 000	**20.** 45 000

B.

1. 6 million **2.** four hundred thousand **3.** thirty thousand **4.** seven thousand **5.** six hundred **6.** forty **7.** eight **8.** seven million four hundred thousand **9.** sixty-five thousand **10.** nine thousand nine hundred and fifty **11.** seventy-six thousand eight hundred **12.** seven million one thousand and thirty-six **13.** four million **14.** three hundred and fifty thousand **15.** nine hundred and forty-five thousand **16.** fifteen thousand **17.** six thousand seven hundred and fifty-three **18.** thirty-four thousand nine hundred and eighty-seven

Exercise 2

A.

1. 63	**6.** 414	**11.** 1118	**16.** 14
2. 152	**7.** 332	**12.** 1430	**17.** 141
3. 264	**8.** 13 936	**13.** 3151	**18.** 23 331
4. 1540	**9.** 608	**14.** 47 931	**19.** 371 600
5. 259	**10.** 71	**15.** 55 746	**20.** 2 149 000

B.

1. 424	**3.** 418	**5.** 5 835 000	**7.** 43
2. 18	**4.** 55 608	**6.** 129	**8.** 803

More sharing (pages 12-14)
Exercise 1

1. 8	**(b)** £2	**(e)** £1.20
2. 4	**(c)** 40	**(f)** £5
3.(a) 3	**(d)** £0.40	**(g)** £10

Exercise 2

1. 10,40	**5.** 70p,£2.10,$\frac{1}{3}$	**9.** 90p,£4.50,£1.80
2. 11,22	**6.** 40p,£1.20,$\frac{3}{4}$	**10.** 12,9,3,8,4,$\frac{1}{9}$
3. 3,15	**7.** 4,6,20,18	
4. 3,15	**8.** 20,10,6,24	

Get the point (pages 15-17)
Exercise 1

2. £0.50	**5.** £0.35	**8.** £0.75	**11.** £0.11
3. £0.30	**6.** £0.90	**9.** £0.15	**12.** £1.35
4. £0.70	**7.** £0.40	**10.** £0.95	**13.** £2.50

Exercise 2

1. £1.90	**5.** £1.76	**9.** £0.26	**13.** £4.55
2. £1.22	**6.** £2.11	**10.** £1.24	**14.(a)** £5.24
3. £3.95	**7.** £22.92	**11.** £2.71	**(b)** 26p
4. £1.82	**8.** £1.75	**12.** £2.31	**(c)** £1.76

Exercise 3

1. £17.75	**5.** £0.30	**9.** £1.06	**13.** £2.35
2. £7.20	**6.** £1.08	**10.** £25.40	**14.** 8p
3. £0.55	**7.** £4.48	**11.** £5.20	**15.** 2 packs of 3
4. £3	**8.** £0.34	**12.** £11.25, £8.75	

Latin and Greek (pages 18-19)
Exercise 1

(a) quadrilateral	**(c)** triangle	**(e)** hexagon
(b) pentagon	**(d)** octagon	**(f)** decagon

Is it enough? (pages 20-22)
Exercise 1

1. yes	**3.** no	**5.** no	**7.** yes	**9.** yes
2. no	**4.** yes	**6.** no	**8.** no	**10.** yes

Exercise 2

1. 5	**3.** 7	**5.** 3 pounds	**7.** 4	**9.** 5
2. 3	**4.** 9	**6.** 4	**8.** 4	**10.** 2 pairs

Check-up 1 (page 23)

1.(a) 51	**(b)** 14	**(c)** 141	**(d)** 18

3. 1208 **4.** thirty-five thousand and fifty six

5. 3,9	**6.(a)** £4.45	**6.(b)** £1.35	**7.** £14.25

8.(a) triangle **(b)** quadrilateral **(c)** hexagon **9.** no **10.** 4 pairs

24 hours in a day (pages 24-25)
Exercise 1

A.

1. 03.00	**4.** 24.00 or 00.00	**7.** 15.00	**10.** 14.00	**13.** 04.00
2. 16.00	**5.** 01.00	**8.** 05.00	**11.** 23.00	**14.** 21.00
3. 12.00	**6.** 19.00	**9.** 18.00	**12.** 08.00	**15.** 22.00

B.

1. 5.00 a.m.	**5.** 1.00 a.m.	**9.** 10.00 p.m.	**13.** 5.00 p.m.
2. 3.00 p.m.	**6.** 4.00 a.m.	**10.** 11.00 a.m.	**14.** 7.00 p.m.
3. midday	**7.** 6.00 a.m.	**11.** 1.00 p.m.	**15.** 8.00 a.m.
4. 8.00 p.m.	**8.** 6.00 p.m.	**12.** 10.00 a.m.	

C.

1. no   **2.** no   **3.** yes

Exercise 2

1. 06.15	**4.** 03.15	**7.** 07.15	**10.** 23.15	**13.** 15.30
2. 12.30	**5.** 16.30	**8.** 07.45	**11.** 18.45	**14.** 21.15
3. 21.45	**6.** 05.30	**9.** 16.30	**12.** 23.45	**15.** 20.45

Exercise 3

1. 07.50	**4.** 12.10	**7.** 20.55	**10.** 23.50	**13.** 02.55
2. 21.20	**5.** 03.35	**8.** 22.35	**11.** 09.10	**14.** 21.50
3. 00.40	**6.** 06.05	**9.** 18.20	**12.** 20.20	**15.** 14.35

Problems (page 27)
Exercise 1

1. £16	**3.** 80	**5.** £1.08	**7.** 48	**9.** 150 m
2. £5	**4.** £4.60	**6.** 14p	**8.** 17, 1	

Centimetres and millimetres (pages 28-29)
Exercise 1

2. 4 cm, 4.0 cm	**8.** 4 cm 5 mm, 4.5 cm
3. 4 cm 8 mm, 4.8 cm	**9.** 1 cm 9 mm, 1.9 cm
4. 9 cm 6 mm, 9.6 cm	**10.** 12 cm 2 mm, 12.2 cm
5. 8 mm, 0.8 cm	**11.** 2 cm 7 mm, 2.7 cm
6. 4 cm 7 mm, 4.7 cm	**12.** 4 mm, 0.4 cm
7. 1 cm 2 mm, 1.2 cm	

Exercise 2

2. 1.0 cm, 10 mm	**7.** 6.4 cm, 64 mm	**12.** 14.4 cm, 144 mm
3. 3.2 cm, 32 mm	**8.** 0.8 cm, 8 mm	**13.** 6.0 cm, 60 mm
4. 4.5 cm, 45 mm	**9.** 3.8 cm, 38 mm	**14.** 9.5 cm, 95 mm
5. 10 cm, 100 mm	**10.** 0.2 cm, 2 mm	**15.** 7.0 cm, 70 mm
6. 3.1 cm, 31 mm	**11.** 7.3 cm, 73 mm	

Moving columns (pages 30-33)
Exercise 1

1. 120	**5.** 390	**9.** 1230	**13.** 4 200 000
2. 50	**6.** 1360	**10.** 4160	**14.** 270 000
3. 150	**7.** 7800	**11.** 10 700	**15.** 9990
4. 240	**8.** 9000	**12.** 3 130 040	

Exercise 2
A.

1. 35	**5.** 500	**9.** 134	**13.** 35 005
2. 42	**6.** 78	**10.** 46	**14.** 4300
3. 2	**7.** 780	**11.** 1 000 000	**15.** 500 000
4. 40	**8.** 1100	**12.** 1000	

B.

(a) 400   **(b)** 1000   **(c)** 4000   **(d)** 36 500   **(e)** 1200

Exercise 3

1. 500	**4.** 76 000	**7.** 19 000	**10.** 12 000	**13.** 29 000
2. 3200	**5.** 999 000	**8.** 12 000	**11.** 3000	**14.** 29 000
3. 45 000	**6.** 78 000	**9.** 43 000	**12.** 3000	**15.** 350 000

Exercise 4

1. 1	**3.** 45	**5.** 40	**7.** 670	**9.** 70 000
2. 35	**4.** 87	**6.** 400	**8.** 7000	

Exercise 5

1. 360 cm	**3.** 12 cm	**5.(a)** 40 in	**6.** 17 500 bricks
2. 4000 mm	**4.** 45 km	**(b)** 400 in	**7.** 35 posts

Directions (pages 34-36)
Exercise 1

1.(a) E	**(b)** W	**(c)** SW	**(d)** S	**(e)** SE	**(f)** N
2.(a) W	**(b)** E	**(c)** SW	**(d)** W	**(e)** NE	**(f)** SW
3.(a) N	**(b)** SE	**(c)** S	**(d)** SW	**(e)** W	**(f)** S

Exercise 2

1. park	**6.** church	**11.** church	**16.** park
2. bridge	**7.** shops	**12.** park	**17.** library
3. Health Centre	**8.** shops	**13.** shops	**18.** bridge
4. church	**9.** library	**14.** youth club	
5. Health Centre	**10.** shops	**15.** church	

Exercise 3

1.(a) S	**(b)** E	**(c)** W	**(d)** N	**(e)** SE
2. running track	**5.** memorial	**8.** tennis courts		
3. pavilion gate	**6.** tennis courts			
4. lido gate	**7.** rose garden			

Calculators (page 37)
Exercise 1
B.

1.

`C` `3` `+` `6` `+` `1` `7` `=`

2.

`C` `9` `×` `2` `÷` `3` `=`

3.

`C` `6` `×` `1` `8` `×` `3` `0` `=`

4.

`C` `5` `1` `+` `1` `7` `+` `1` `3` `4` `=`

5.

`C` `2` `9` `8` `−` `1` `4` `+` `6` `3` `=`

6.

`C` `1` `0` `2` `3` `÷` `3` `=`

Columns for multiplying (pages 37-38)
Exercise 1

1. 114	**4.** 486	**7.** 192	**10.** 112	**13.** 232
2. 120	**5.** 72	**9.** 216	**11.** 51	**14.** 148
3. 111	**6.** 132	**9.** 81	**12.** 450	**15.** 495

Exercise 2

2. 300	**8.** 1200	**14.** 13 500	**20.** 750
3. 1050	**9.** 13 500	**15.** 4250	**21.** 520
4. 450	**10.** 6650	**16.** 19 800	**22.** 12 500 g
5. 9000	**11.** 1350	**17.** 600	**23.** 3750 g
6. 2500	**12.** 25 600	**18.** 420	
7. 1000	**13.** 24 500	**19.** 3600	

More than a whole one (pages 40-41)
Exercise 1

1. $1\frac{1}{2}$	**3.** $1\frac{2}{3}$	**5.** $2\frac{1}{8}$	**7.** $2\frac{3}{4}$
2. $1\frac{3}{4}$	**4.** $1\frac{4}{5}$	**6.** $3\frac{3}{10}$	

Exercise 2

1. 9	**2.** 11	**3.** 7	**4.** 15	**5.** 12	**6.** 8

Metres and centimetres (page 43)
Exercise 1
A.
1. 1.64 m **2.** 3.84 m **3.** 2.5 m **4.** 1.96 m **5.** 1.1 m
B.
1. 265 cm **2.** 663 cm **3.** 1800 cm **4.** 116 cm **5.** 12 cm

Check-up 2 (page 44)
1.(a) 08.00 **(b)** 16.30 **(c)** 21.50
2.(a) 7.30 a.m. **(b)** 5.20 p.m. **(c)** 2.50 p.m.
3. £1.20 **4.(a)** 1.4 cm **(b)** 3.8 cm **5.(a)** 23 mm **(b)** 40 mm
6.(a) 140 **(b)** 1600 **(c)** 36 **(d)** 45
7.(a) NE **(b)(i)** Homeley **(ii)** Crickley
8.(a) 72 **(b)** 450 **(c)** 3400 **9.** $3\frac{4}{5}$ **10.** 15
11.(a) 3.12 m **(b)** 2.18 m **(c)** 0.46 m

Factors and primes (page 45)
Exercise 2
11, 29, 19, 47 67 are prime

Shorthand (page 84)
Exercise 1
A.
1. 3p **4.** 5q **7.** 3st
2. 4b **5.** 6r **8.** 35a
3. pq **6.** 4rs **9.** 2ab
B.
1. $2 \times a$ **4.** $13 \times p$ **7.** $6 \times p \times q$
2. $3 \times b$ **5.** $p \times r$ **8.** $18 \times r \times s$
3. $4 \times r$ **6.** $8 \times a$ **9.** $4 \times a \times b \times c$
Exercise 2
1. $a/3$ **4.** $4/3p$ **7.** $3p/5r$
2. $2/b$ **5.** p/q **8.** pr/ab
3. $2a/3$ **6.** $2b/p$ **9.** ab/pqr
Exercise 3
1. 3a **4.** 3p **7.** 4l
2. 2b **5.** 5a **8.** 6c
3. 4p **6.** 2r **9.** 3d
Exercise 4
1. 5a **4.** 10p **7.** 20y
2. 9b **5.** 5m **8.** 7r
3. 8b **6.** 10m **9.** 7r
Exercise 5
1. p **4.** m **7.** 3p
2. 3a **5.** 3m **8.** 4m
3. 3a **6.** 7r **9.** 8y

Division sums (pages 48-49)
Exercise 1
1. 0.25 **4.** 0.2 **7.** 0.375 **10.** 0.625
2. 0.5 **5.** 0.8 **8.** 0.333... **11.** 0.0625
3. 0.1 **6.** 0.7 **9.** 0.666... **12.** 0.111...
Exercise 2
1. 1.5 **3.** 3.4 **5.** 6.375 **7.** 4.33′ **9.** 3.7
2. 2.25 **4.** 1.75 **6.** 2.1 **8.** 2.625 **10.** 4.6

Special quadrilaterals (page 51)
Exercise 1
1. square **6.** square **11.** trapezium
2. parallelogram **7.** trapezium **12.** rectangle
3. rhombus **8.** trapezium **13.** rhombus
4. rectangle **9.** parallelogram
5. parallelogram **10.** kite

The same thing (pages 52-55)
Exercise 1
1. $2/8 = 1/4$ **5.** $4/16 = 1/4$ **9.** $10/16 = 5/8$
2. $4/8 = 1/2$ **6.** $12/16 = 3/4$ **10.** $14/16 = 7/8$
3. $6/8 = 3/4$ **7.** $2/16 = 1/8$
4. $8/16 = 1/2$ **8.** $6/16 = 3/8$
Exercise 2
1. $2/6 = 1/3$ **6.** $4/12 = 1/3$ **11.** $3/9 = 2/6$
2. $3/9 = 1/3$ **7.** $2/12 = 1/6$ **12.** $3/6 = 1/2$
3. $4/6 = 2/3$ **8.** $8/12 = 2/3$ **13.** $6/12 = 1/2$
4. $6/9 = 2/3$ **9.** $4/12 = 3/9$ **14.** $3/12 = 1/4$
5. $6/9 = 4/6$ **10.** $8/12 = 4/6$ **15.** $2/10 = 1/5$
Exercise 4
2. 1/2 **6.** 3/4 **10.** 1/2 **14.** 1/4
3. 1/2 **7.** 3/4 **11.** 1/3 **15.** 1/3
4. 1/4 **8.** 1/2 **12.** 1/3 **16.** 2/3
5. 1/3 **9.** 1/3 **13.** 1/4 **17.** 2/3

Time to go (page 59)
Exercise 1
1.(a) 1 hr 15 min **(b)** 2 hr 45 min **(c)** 3 hr 27 min
2.(a) 6 hr 30 min **(b)** 2 hr 45 min **(c)** 3 hr 40 min
3.(a) 1 hr 5 min **(b)** 1 hr 50 min **(c)** 5 hr 10 min
(d) 6 hr 25 min **(e)** 3 hr 15 min **(f)** 1 hr 30 min
4.(a) 10 min **(b)** 1 hr 50 min **(c)** 55 min
(d) 1 hr 25 min **(e)** 2 hr 45 min **(f)** 5 hr 5 min
5.(a) 55 min **(b)** 4 hr 10 min **(c)** 2 hr 35 min
(d) 6 hr 45 min **(e)** 37 min **(f)** 3 hr 30 min

Flow diagrams (pages 60-63)
Exercise 1
1. 1 **2.** 7 **3.** 3 **4.** 17 **5.** 5 **6.** 18 **7.** 1
1.(a) 0 **(b)** 1 **(c)** 3 **4.(a)** 24 **(b)** 44 **(c)** 84
2.(a) 14 **(b)** 19 **(c)** 29 **5.(a)** 20 **(b)** 24 **(c)** 44
3.(a) 9 **(b)** 12 **(c)** 18 **6.(a)** 48 **(b)** 52 **(c)** 72

Angles (pages 64-65)
Exercise 1
A.
1. full **3.** quarter **5.** quarter **7.** half
2. quarter **4.** half **6.** full **8.** quarter
B.
1. quarter clockwise **5.** quarter anticlockwise
2. half clockwise **6.** half clockwise
3. quarter anticlockwise **7.** quarter anticlockwise
4. full clockwise **8.** quarter anticlockwise
Exercise 2
2. a right angle **5.** a right angle **8.** more
3. more **6.** less **9.** less
4. more **7.** more

Dividing (page 66)
Exercise 1
A.
1. 102 **3.** 107 **5.** 15 **7.** 35 **9.** 107
2. 104 **4.** 18 **6.** 19 **8.** 220 **10.** 30

B.
1. 150   **2.** 108   **3.** £86   **4.** £16, £2

Check-up 3 (page 67)
2.(a) 5p   **(b)** 3ab   **3.(a)** 9p   **(b)** 5b
4.(a) 0.75   **(b)** 0.8   **(c)** 2.7
5.(a) [rectangle]   **(b)** [parallelogram]   **(c)** [rhombus]
6.(a) ¾   **(b)** ½   **(c)** ½   **7.** 5
8.(a) more   **(b)** less   **(c)** less
9.(a) 25 min   **(b)** 1 hr 5 min   **(c)** 2 hr 40 min
10.(a) 102   **(b)** 108   **(c)** 403

Adding and substracting fractions (pages 68-70)
Exercise 1
1. ³/₈   **2.** ³/₃   **3.** ⁴/₅   **4.** ⁴/₈   **5.** ⁸/₁₀   **6.** ⁴/₃

Exercise 2.
2. ³/₄   **5.** ⁵/₈   **8.** ¹¹/₈
3. ⁵/₁₀   **6.** ⁹/₁₀   **9.** ¹¹/₁₀
4. ⁷/₆   **7.** ¹¹/₁₆   **10.** ¹⁰/₉

Exercise 3
1. ¹¹/₁₂   **2.** ⁵/₆   **3.** ¹³/₂₀   **4.** ¹¹/₁₂   **5.** 1³/₁₀   **6.** 1⁴/₁₅

Exercise 4
1. ¼   **4.** ⅛   **7.** ¹/₁₀
2. ³/₈   **5.** ¹/₁₅   **8.** ⅛
3. ⅓   **5.** ⅓   **9.** ⅙

Decimals (page 72)
Exercise 1
1. 36   **4.** 1.9   **7.** 93.3   **10.** 1.96   **13.** 2.21
2. 0.02   **5.** 463   **8.** 4.63   **11.** 345   **14.** 2.21
3. 3.42   **6.** 27.4   **9.** 0.27   **12.** 22.1   **15.** 0.1

Triangles (page 74-75)
Exercise 1 only major name given
1. right angled   **6.** equilateral   **11.** equilateral
2. isosceles   **7.** isosceles   **12.** isosceles
3. isosceles   **8.** equilateral   **13.** isosceles
4. acute angled   **9.** isosceles   **14.** acute angled
5. obtuse angled   **10.** acute angled   **15.** right angled

Surveys (pages 76-81)
Exercise 1
1.

vehicle	tally	number
cars	̶I̶I̶I̶I̶ ̶I̶I̶I̶I̶ ̶I̶I̶I̶I̶ ̶I̶I̶I̶I̶ ̶I̶I̶I̶I̶ II	27
lorries	III	3
buses	̶I̶I̶I̶I̶ IIII	9
taxis	I	1
vans	̶I̶I̶I̶I̶ I	6
motorcycles	̶I̶I̶I̶I̶ III	8
bicycles	̶I̶I̶I̶I̶ ̶I̶I̶I̶I̶ III	13
	total:	67

2.

method	tally	number
car	III	3
bus	̶I̶I̶I̶I̶ II	7
train	I	1
bicycle	̶I̶I̶I̶I̶ III	8
walk	̶I̶I̶I̶I̶ ̶I̶I̶I̶I̶ I	11
	total:	30

Exercise 2
1.

score	tally	number
1	II	2
2	III	3
3	̶I̶I̶I̶I̶	5
4	IIII	4
5	II	2
6	IIII	4
7		0
8	II	2
9	IIII	4
10	II	2
	total:	28

2.

size	tally	number
2	̶I̶I̶I̶I̶	5
3	̶I̶I̶I̶I̶ II	.7
4	̶I̶I̶I̶I̶ III	8
5	̶I̶I̶I̶I̶ I	6
6	III	3
7	II	2
8		0
9	I	1
	total:	32

Most common: 4   Least common: 8

Exercise 3
1.

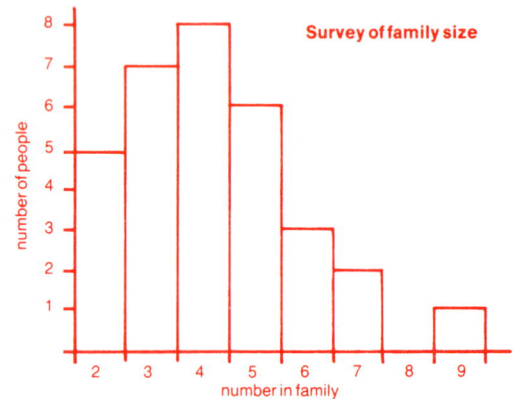

Survey of family size

2.(a) 6   **(b)** 4   **(c)** 3   **(d)** 4

Exercise 4
1.

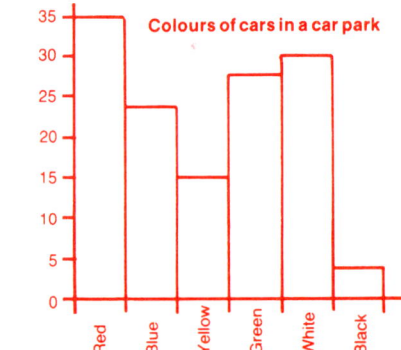

Colours of cars in a car park

182

2.

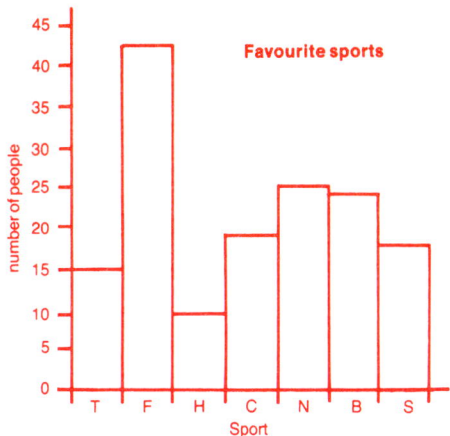

Favourite sports (bar chart: number of people vs Sport — T, F, H, C, N, B, S)

3.(a) April   **(b)** June and July   **(c)** 12 cm
(d) 7 cm   **(e)** 6 cm   **(f)** 4 cm

4.

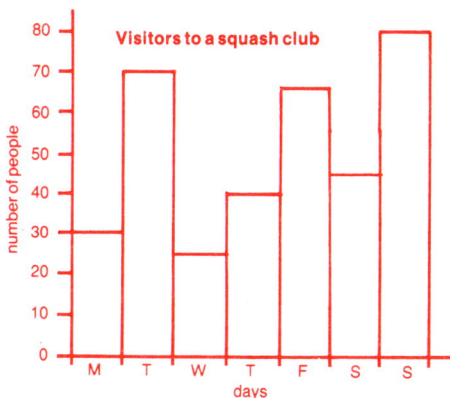

Visitors to a squash club (bar chart: number of people vs days — M, T, W, T, F, S, S)

5.(a) 0.4 million   **(b)** 0.5 million   **(c)** 0.3 million
(d) by 0.2 million   **(e)** 4 million

6.(a)

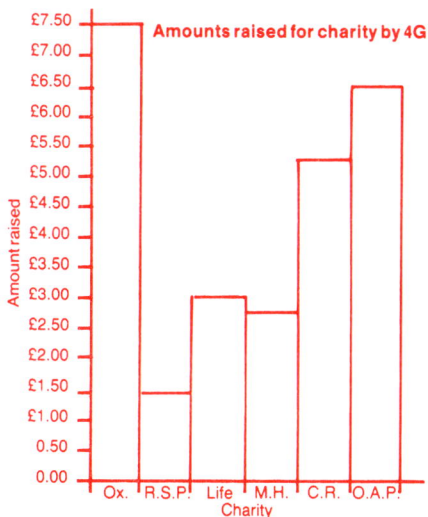

Amounts raised for charity by 4G (bar chart: Amount raised vs Charity — Ox., R.S.P., Life, M.H., C.R., O.A.P.)

(b) £26.50 raised altogether

Which units? (pages 82-83)
Exercise 1

1. cm	**3.** mm	**5.** m	**7.** m	**9.** mm
2. km	**4.** m	**6.** km	**8.** mm	**10.** cm

Exercise 2

1. 2	**7.** 5800	**13.** 3.5	**19.** 1600	**25.** 65
2. 1.5	**8.** 6421	**14.** 4.25	**20.** 62	**26.** 85
3. 50	**9.** 10 324	**15.** 146	**21.** 620	
4. 3.512	**10.** 2230	**16.** 35	**22.** 1340	
5. 4.036	**11.** 18	**17.** 0.12	**23.** 18 400	
6. 4000	**12.** 2.5	**18.** 3.67	**24.** 3.5	

Splitting it up (page 84)
Exercise 1

2. 3375	**4.** 9375	**6.** 54 625	**8.** 78 480
3. 24 750	**5.** 19 375	**7.** 18 480	**9.** 58 125

Time means of (pages 86-87)
Exercise 1

2. $\frac{1}{4}$	**4.** $\frac{1}{6}$	**6.** $\frac{1}{16}$	**8.** $\frac{1}{8}$	**10.** $\frac{1}{6}$
3. $\frac{1}{8}$	**5.** $\frac{1}{6}$	**7.** $\frac{1}{7}$	**9.** $\frac{1}{3}$	**11.** $\frac{1}{12}$

Exercise 2

1. $\frac{1}{12}$	**3.** $\frac{3}{8}$	**5.** $\frac{1}{10}$	**7.** $\frac{5}{9}$	**9.** $\frac{3}{40}$
2. $\frac{2}{15}$	**4.** $\frac{1}{32}$	**6.** $\frac{1}{9}$	**8.** $\frac{3}{10}$	**10.** $\frac{1}{18}$

Variable inputs (pages 88-90)
Exercise 1
A.

1. 4 + 8 = 12	**4.** 8 + 63 = 71	**7.** 96 + 18 = 114
2. 3 + 6 = 9	**5.** 19 + 107 = 126	**8.** 45 + 53 = 98
3. 17 + 4 = 21	**6.** 1000 + 101 = 1101	**9.** 136 + 27 = 163

B.

1. 3, 5, 8	**4.** 47, 16, 63	**7.** 11, 11, 22
2. 4, 9, 13	**5.** 27, 148, 175	**8.** 39, 41, 80
3. 120, 35, 155	**6.** 21, 22, 43	**9.** 27, 127, 154

Exercise 2
A.

1. 9 − 4 = 5	**4.** 36 − 18 = 18	**7.** 883 − 57 = 826
2. 16 − 7 = 9	**5.** 26 − 4 = 22	**8.** 360 − 210 = 150
3. 114 − 27 = 87	**6.** 450 − 97 = 353	**9.** 115 − 115 = 0

B.

1. 18, 6, 12	**4.** 63, 45, 18	**7.** 590, 286, 304
2. 44, 12, 32	**5.** 71, 18, 53	**8.** 312, 147, 165
3. 39, 15, 24	**6.** 46, 27, 19	**9.** 205, 32, 173

Exercise 3
A.

1. 7	**2.** 14	**3.** 18	**4.** 106	**5.** 34	**6.** 59

B.

1. 17	**2.** 8	**3.** 24	**4.** 80	**5.** 72	**6.** 63

C.

1. 2	**2.** 2	**3.** 11	**4.** 5	**5.** 65	**6.** 11

Check-up 4 (page 91)

1.(a) $\frac{1}{2}$	**(b)** $\frac{1}{2}$	**(c)** $\frac{3}{10}$
2.(a) right angled	**(b)** isosceles	**(c)** obtuse angled
3.(a) 48	**(b)** 206	**(c)** 430
4.(a) 490	**(b)** 3750	**5.** 29

6.

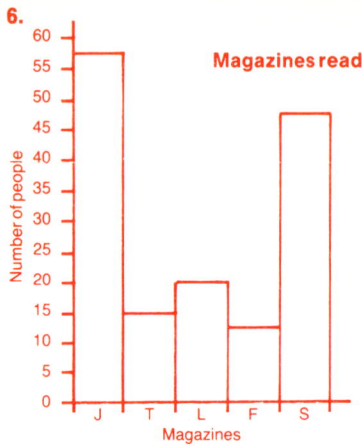

Magazines read (bar chart: y-axis "Number of people" 0-60, x-axis "Magazines" with J, T, L, F, S)

8.(a) 2.5   **(b)** 3650   **9.(a)** $\frac{1}{10}$ **(b)** $\frac{1}{4}$ **(c)** $\frac{5}{18}$
10.(a) 5   **(b)** 2

Multiplying and dividing decimals (page 92)
Exercise 1
1. 7.5   **4.** 0.61   **7.** 41.03   **10.** 56p   **13.** 0.03 kg
2. 15   **5.** 6.5   **8.** 35.4   **11.** 6 kg   **14.** 0.18 m
3. 0.25   **6.** 18   **9.** 7.92   **12.** 592.5 kg **15.** 2.7 g

How many? (pages 93-95)
Exercise 1
2. 3   **3.** 2   **4.** 2   **5.** 6   **6.** 4   **7.** 9

Exercise 2
2. $\frac{3}{4}$ **4.** 4   **6.** $\frac{2}{5}$ **8.** $\frac{5}{9}$ **10.** $1\frac{2}{3}$ **12.** $1\frac{1}{5}$
3. $1\frac{1}{2}$ **5.** $1\frac{1}{8}$ **7.** $\frac{9}{10}$ **9.** $1\frac{4}{5}$ **11.** $2\frac{4}{5}$ **13.** $\frac{5}{16}$

Exercise 3
1. $\frac{2}{5}$ **4.** 5, 4, 2, 9   **7.** $\frac{7}{8}$ lb
2. $1\frac{1}{8}$ lb   **5.** £2, 50p, 50p, $\frac{1}{6}$ **8.** 6, 8, 3, 2, 5
3. £113.25   **6.** £9.75

Areas (pages 96-97)
1. 12 cm²   **4.** 15 cm²   **7.** 20 cm²   **10.** 15 cm²   **13.** 6 cm²
2. 10 cm²   **5.** 24 cm²   **8.** 14 cm²   **11.** 24 cm²   **14.** 6 cm²
3. 9 cm²   **6.** 12 cm²   **9.** 6 cm²   **12.** 7 cm²   **15.** 4 cm²

Grams and kilograms (page 98)
Exercise 1
1. 2   **5.** 0.215   **9.** 3500   **13.** 60
2. 2.5   **6.** 0.12   **10.** 4200   **14.** 543
3. 6   **7.** 0.035   **11.** 6305   **15.** 18
4. 4.78   **8.** 0.28   **10.** 500   **16.** 342

Missing numbers (page 99)
Exercise 1
1. 4   **5.** 2   **9.** 13   **13.** 30   **17.** 3
2. 1   **6.** 3   **10.** 53   **14.** 18   **18.** 12
3. 2   **7.** 2   **11.** 20   **15.** 12   **19.** 7
4. 3   **8.** 12   **12.** 6   **16.** 12   **20.** 24

Calculator problems (pages 100-101)
Exercise 1
1. £52.34, £2.66   **2.** 2660   **3.** 1296

4. £83   **5.** £4082   **6.** A by £104 per year
7. 50   **8.** £12.35, £321.10
9.(a) £1.97   **(b)** £3.15   **(c)** £3.44   **(d)** £6.17   **(e)** £4.29

Co-ordinates (pages 102-103)
Exercise 1
A.
1.(a) pond   **(c)** tower   **(e)** secret tunnel
   (b) pirate ship   **(d)** pirate camp   **(e)** lighthouse
2.(a) (2,8)   **(c)** (1,3)   **(e)** (9,3)
   (b) (8,2)   **(d)** (7,4)   **(e)** (7,7½)

B.
1.(a) bridge   **(d)** Rose Cottage   **(g)** bus stop
   (b) Manor Farm   **(e)** station   **(h)** signals
   (c) Manor Cottages   **(f)** Glebe Farm   **(i)** teachers house
2.(a) (4,3)   **(d)** (9,5)   **(g)** (8½,4½)
   (b) (4,8)   **(e)** (2½,8)   **(h)** (9½,2½)
   (c) (4,7)   **(f)** (2½,5½)   **(i)** (8,7)

Pictograms (pages 104-105)
Exercise 1
1. 400   **2.** 250   **3.** 200   **4.** 50   **5.** 100

Exercise 2
1.

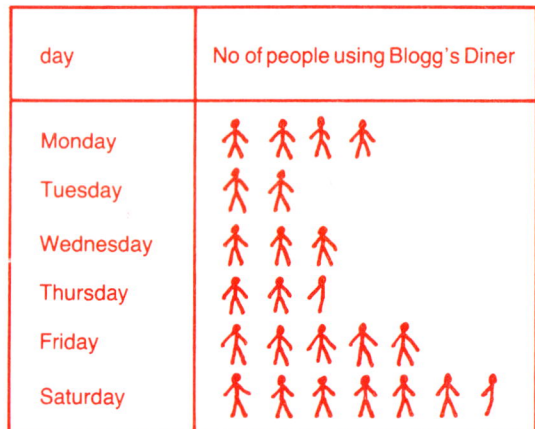

day	No of people using Blogg's Diner
Monday	🧍🧍🧍🧍
Tuesday	🧍🧍
Wednesday	🧍🧍🧍
Thursday	🧍🧍🧍
Friday	🧍🧍🧍🧍🧍
Saturday	🧍🧍🧍🧍🧍🧍🧍

🧍 represents 10 people

2.
(a) 60   **(b)** 30   **(c)** 25   **(d)** 75   **(e)** 25   **(f)** 15
3.

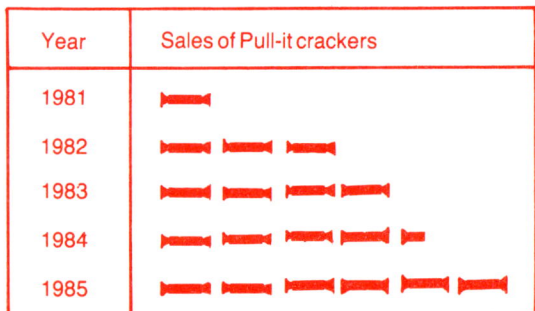

Year	Sales of Pull-it crackers
1981	▰
1982	▰ ▰ ▰
1983	▰ ▰ ▰ ▰
1984	▰ ▰ ▰ ▰ ▰
1985	▰ ▰ ▰ ▰ ▰ ▰

▰ represents 20 000 boxes

Rule of zeros (pages 106-107)
Exercise 1

1. 80	**6.** 36 000	**11.** 25 000	**16.** 5500
2. 1500	**7.** 300	**12.** 6000	**17.** 2100
3. 8000	**8.** 160	**13.** 100 000	**18.** 9600
4. 600	**9.** 4200	**14.** 1800	**19.** 3600
5. 3600	**10.** 12 000	**15.** 480	**20.** 660 000

Exercise 2

1. ✓	**6.** ✗	**11.** ✓	**16.** ✓
2. ✓	**7.** ✓	**12.** ✗	**17.** ✓
3. ✗	**8.** ✓	**13.** ✗	**18.** ✗
4. ✓	**9.** ✓	**14.** ✓	**19.** ✗
5. ✗	**10.** ✓	**15.** ✗	**20.** ✓

Degrees (pages 109-111)
Exercise 1

A.

1. d	**3.** j	**5.** e	**7.** h	**9.** i
2. c	**4.** b	**6.** a	**8.** f	**10.** g

B.

1. 85°	**3.** 250°	**5.** 30°	**7.** 45°	**9.** 260°
2. 160°	**4.** 270°	**6.** 345°	**8.** 75°	**10.** 190°

Parallelogram areas (page 113)

2. 30 cm²	**5.** 45 m²	**8.** 20 ft²	**11.** 10 000 m²
3. 36 m²	**6.** 12 cm²	**9.** 28 mm²	
4. 200 m²	**7.** 55 in²	**10.** 200 cm²	

Check-up 5 (page 114)

1.(a) 10.4	**(b)** 261	**(c)** 9.1	**(d)** 0.81
2.(a) 2	**(b)** 1⅕	**(c)** 4	
3.(a) 15 cm²	**(b)** 36 cm²	**4.(a)** 3.5	**(b)** 3640
5.(a) 2	**(b)** 7	**(c)** 18	
6.(a) Farm	**(b)** (3,3)	**(c)** (1,1½)	
7.			

Month	No of rainy days
Jan.	⟁ ⟁ ⟁ ⟁ ⟁ ⟁
Feb.	⟁ ⟁ ⟁ ⟁ ⟁ ⟁ ⟁ ⟁ ⟁
March	⟁ ⟁ ⟁ ⟁ ⟁ ⟁ ⟁ ⟁
April	⟁ ⟁ ⟁ ⟁ ⟁ ⟁ ⟁
May	⟁ ⟁ ⟁ ⟁ ⟁
June	⟁ ⟁ ⟁ (

⟁ represents 2 rainy days

8.(a) 600	**(b)** 200	**(c)** 18 000	**9.(a)** 45°	**(b)** 100°
10.(a) 20 m²	**(b)** 96 ft²			

Out of a hundred (pages 115-118)
Exercise 1

1. 36%	**2.** 21%

Exercise 2
1. 27, 16, 12, 9

185

Exercise 3
1. 86, 44, 26, 24, 14
2. 135, 50, 40, 60, 95, 30
3. 105, 63, 6, 57, 12, 36, 21

Exercise 4
1. 12%, 19%, 6%, 23%, 9%
2. 27%, 32%, 15%, 5%, 18%, 3%
3. 76%, 80%, 72%, 54%, 48%, 30%
4.(a) 30% **(b)** 140 **(c)** 60

Brackets (pages 119-120)
Exercise 1

1. $4 \times 2 + 3 = 11$	**5.** $7 \times 2 + 20 = 34$	**9.** $10 \div 2 + 14 = 19$
2. $10 \times 2 - 3 = 17$	**6.** $5 \times 3 + 2 + 4 = 21$	**10.** $16 \div 4 + 5 = 9$
3. $6 \div 3 + 4 = 6$	**7.** $8 \div 4 - 1 = 1$	**11.** $4 \times 3 + 24 - 8 = 28$
4. $7 \times 4 + 3 = 31$	**8.** $6 \times 3 + 12 = 30$	**12.** $12 \div 3 + 14 - 5 = 13$

Exercise 2

1. 14	**4.** 54	**7.** 10	**10.** 4
2. 36	**5.** 5	**8.** 16	**11.** 81
3. 8	**6.** 10	**9.** 4	**21.** 1

More co-ordinates (pages 121-123)
Exercise 1

1. (6,1) (4,1) (3,3) (5,3) (5,5) (2,4) (1,5) (2,6) (5,5) (2,7) (2,9) (4,8) (5,5) (6,8) (8,9) (8,7) (5,5) (8,6) (9,5) (8,4) (5,5) (5,3) (7,3) (6,1)

2. (9,7) (9,2) (8,1) (6,1) (5,1½) (4,1) (2,1) (1,2) (1,7) (2,8) (4,8) (5,7) (5,10) (6,10) (6,9) (7,10) (8,10) (8,9) (6,9) (5,7) (6,8) (8,8) (9,7)

3. (5,10) (8,8) (6,8) (8,6) (6,6) (9,4) (6,4) (6,3) (7,3) (6,1) (4,1) (3,3) (4,3) (4,4) (1,4) (4,6) (2,6) (4,8) (2,8) (5,10)

4. (8,10) (10,7) (7,7) (6,8) (4,8) (4,7) (5,6) (8,6) (10,5) (10,2) (8,1) (2,1) (1,3) (4,3) (5,2) (6,2) (7,3) (7,4) (5,5) (2,5) (1,6) (1,8) (3,10) (8,10)

5. (2,2) (2,5) (1,5) (1,6) (2,6) (2,7) (3,7½) (5,7½) (6,7) (6,6) (7,6) (7,5) (6,5) (7,3) (6,3) (6,2) (5,2) (6,1) (4,1) (2,2)

6. (8,4) (4,2) (2,3) (2,7) (½,4) (½,7½) (2½,7½) (7½,10) (4½,10) (3,8) (6,8) (6,10) (8,10) (10,8) (8,8)

Exercise 2
1.

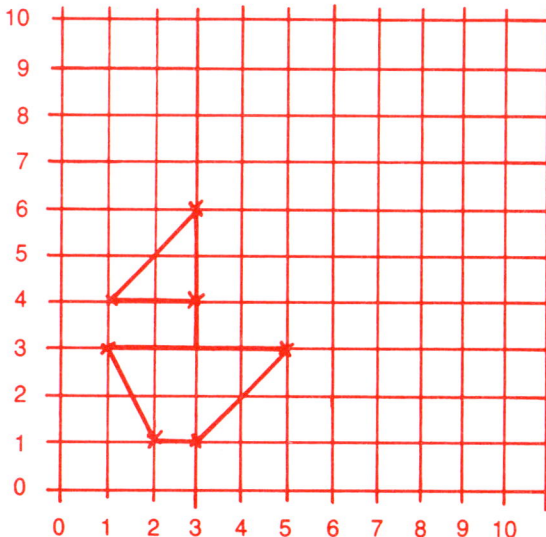

2.

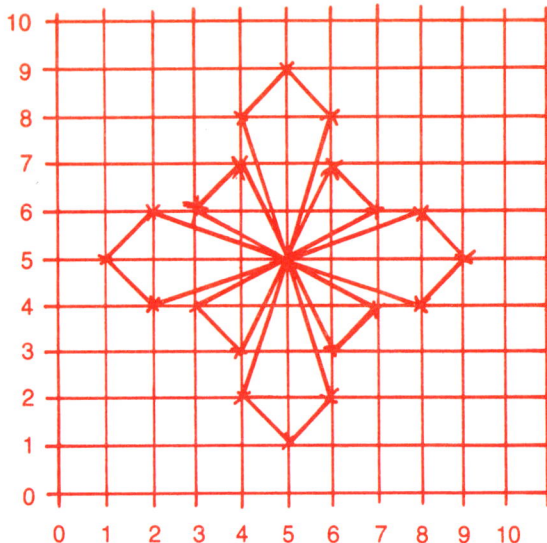

3.

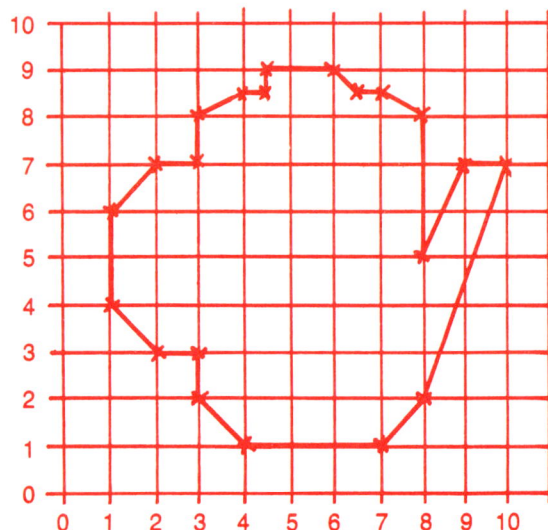

Triangle areas (page 125)
Exercise 1
2. 10 cm² **5.** 24½ mm³ **8.** 15 ft² **11.** 8 cm²
3. 20 cm² **6.** 33 in² **9.** 54 cm²
4. 6 m² **7.** 75 ft² **10.** 100 mm²

Millilitres and litres (page 126)
Exercise 1
1. 4 **5.** 2.196 **9.** 625 **13.** 750
2. 2.3 **6.** 0.025 **10.** 3000 **14.** 36 400
3. 0.45 **7.** 27 **11.** 436 **15.** 50
4. 0.05 **8.** 46.005 **12.** 2710 **16.** 3750

1p in the £1 (page 127)
Exercise 1
1. 5p **7.** 75p **13.** 30p **19.** £1
2. 20p **8.** 60p **14.** 12p **20.** 50p
3. 30p **9.** 47p **15.** 80p **21.** 30p
4. 50p **10.** 80p **16.** 50p **22.** £3
5. 45p **11.** £1 **17.** £1.50 **23.** £4.50
6. 90p **12.** 15p **18.** £1 **24.** 90p
25. 80p, £7.20 **26.** £9 **27.** £15, £115
28.(a) £12 **(b)** £45 **(c)** £1.50 **(d)** 75p **(e)** £33

How much do I use? (pages 128-129)
Exercise 1
1. 100 g **3.** 400 g **5.** 300 g **7.** 700 g **9.** 1500 g
2. 50 g **4.** 600 g **6.** 1000 g **8.** 2000 g

Exercise 2
1. 100 g **3.** 150 g **5.** 50 g **7.** 500 g **9.** 375 g
2. 300 g **4.** 25 g **6.** 250 g **8.** 400 g

Exercise 3
1.(a) 4 teasp **(b)** 12 teasp **(c)** 1 teasp
2.(a) 3 teasp **(b)** 6 teasp **(c)** 1½ teasp
3.(a) 1 teasp **(b)** 2 teasp **(c)** ½ teasp
4.(a) 1 teasp **(b)** 3 teasp **(c)** 1½ teasp

Measuring angles (pages 131-132)
Exercise 1
1. 21° **4.** 50° **7.** 108° **10.** 12°
2. 60° **5.** 45° **8.** 156° **11.** 90°
3. 132° **6.** 35° **9.** 280° **12.** 238°

Yards feet and inches (page 133)
Exercise 1
1.(a) 24 **(b)** 60 **(c)** 36 **(d)** 18 **(e)** 48 **(f)** 72
2.(a) 9 **(b)** 6 **(c)** 1½ **(d)** 12 **(e)** 30 **(f)** 15
3.(a) 3 **(b)** 2 **(c)** 5 **(d)** 4 **(e)** 6 **(f)** 1½
4.(a) 40p **(b)** 60p **(c)** £1.80 **(d)** 30p **(e)** 90p **(f)** £4.50
5.(a) 1 **(b)** 2 **(c)** 3 **(d)** 6 **(e)** 1½ **(f)** ½

Check-up 6 (page 134)
1. 92%, 74%, 84%
2.(a) 30 **(b)** 3 **3.(a)** 9 cm² **(b)** 10 cm²
4.(a) 2 **(b)** 4600 **5.(a)** 10p **(b)** £3 **(c)** 10p
6.(a) 4p **(b)** 16p **(c)** 24p
7. You should get a T shape
8.(a) 1½ teasp **(b)** 6 teasp **9.(a)** 95° **(b)** 30° **(c)** 315°

Percentages to fractions (page 135)
Exercise 1
2. ½ **6.** ¾ **10.** ⁹/₁₀ **14.** ¹/₁₀₀
3. ¹/₁₀ **7.** ³/₅ **11.** ⁹/₂₀ **15.** ¹/₂₀
4. ⅕ **8.** ⁷/₁₀ **12.** 1
5. ³/₁₀ **9.** ⅘ **13.** ¹⁹/₂₀
16.(a) ¼ **(b)** 75° **(c)** ¾
17. ½ **18.(a)** ¾ **(b)** ¼

Perimeters (pages 136-137)
Exercise 1
1. 12 cm **3.** 18 m **5.** 36 mm
2. 20 cm **4.** 80 m **6.** 80 cm

Exercise 2

1. 28 cm   **3.** 70 m   **5.** 35 cm   **7.** 32 m
2. 26 m   **4.** 60 m   **6.** 84 cm   **8.** 22 cm

Changes (pages 138-140)
Exercise 1

2.(a) 2nd Sun   **(b)** 1st Sun   **(c)** 1st Tue-Sun
3.

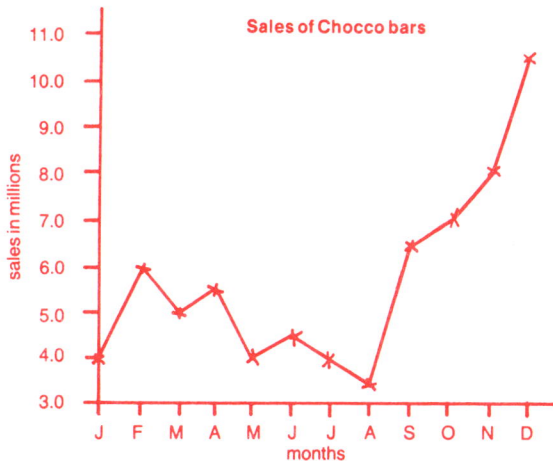

4.(a) every 2 hours   **(b)** 40°   **(c)** 37.5°
(d) 20.00   **(e)** 08.00
5.

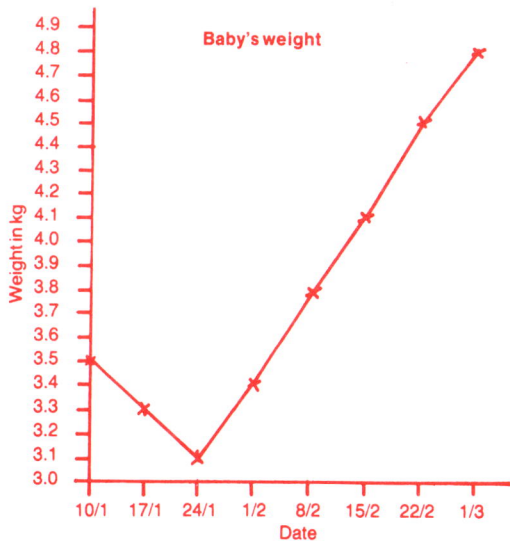

Decimals to fractions (page 141)
Exercise 1

1. ½   **5.** ¼   **9.** ¹/₂₀   **13.** ³/₂₀
2. ⅕   **6.** ¾   **10.** ¹/₁₀₀   **14.** ⅖
3. ⅗   **7.** ³/₁₀   **11.** ⅘   **15.** ¹⁹/₂₀
4. ⁹/₁₀   **8.** ⁷/₂₀   **12.** ¹⁷/₂₀   **16.** ²³/₅₀

More flow diagrams (pages 142-145)
Exercise 2

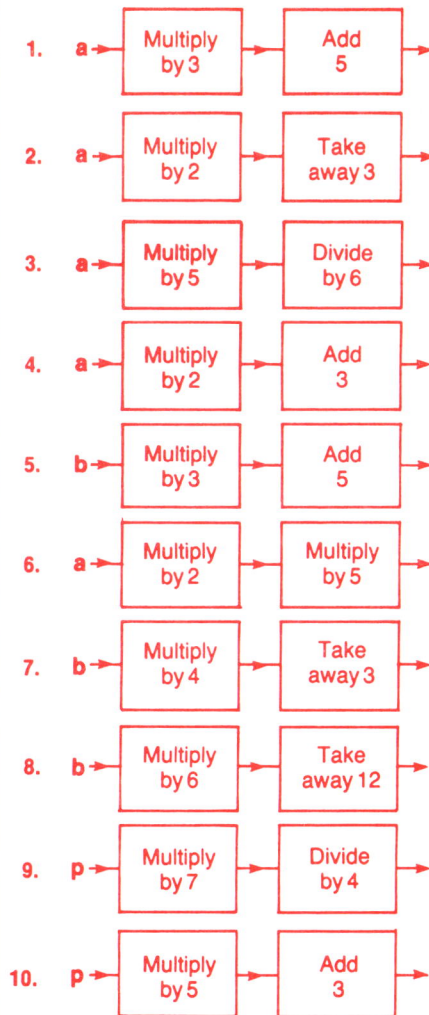

Exercise 3

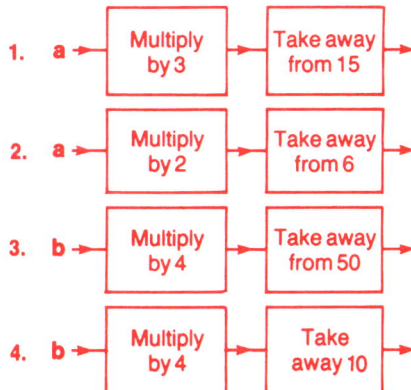

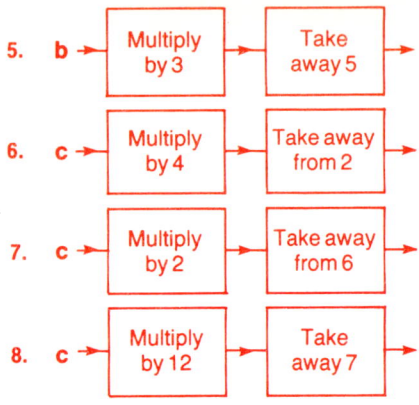

5. **b** → | Multiply by 3 | → | Take away 5 | →

6. **c** → | Multiply by 4 | → | Take away from 2 | →

7. **c** → | Multiply by 2 | → | Take away from 6 | →

8. **c** → | Multiply by 12 | → | Take away 7 | →

Exercise 4

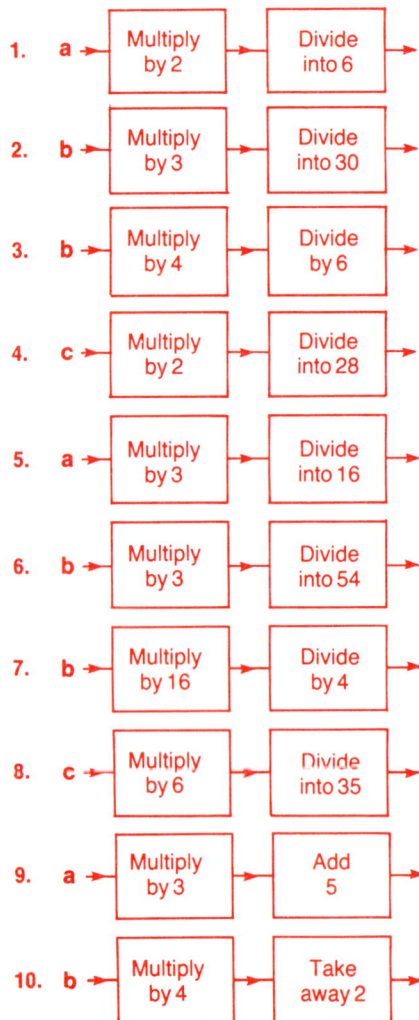

1. **a** → | Multiply by 2 | → | Divide into 6 | →

2. **b** → | Multiply by 3 | → | Divide into 30 | →

3. **b** → | Multiply by 4 | → | Divide by 6 | →

4. **c** → | Multiply by 2 | → | Divide into 28 | →

5. **a** → | Multiply by 3 | → | Divide into 16 | →

6. **b** → | Multiply by 3 | → | Divide into 54 | →

7. **b** → | Multiply by 16 | → | Divide by 4 | →

8. **c** → | Multiply by 6 | → | Divide into 35 | →

9. **a** → | Multiply by 3 | → | Add 5 | →

10. **b** → | Multiply by 4 | → | Take away 2 | →

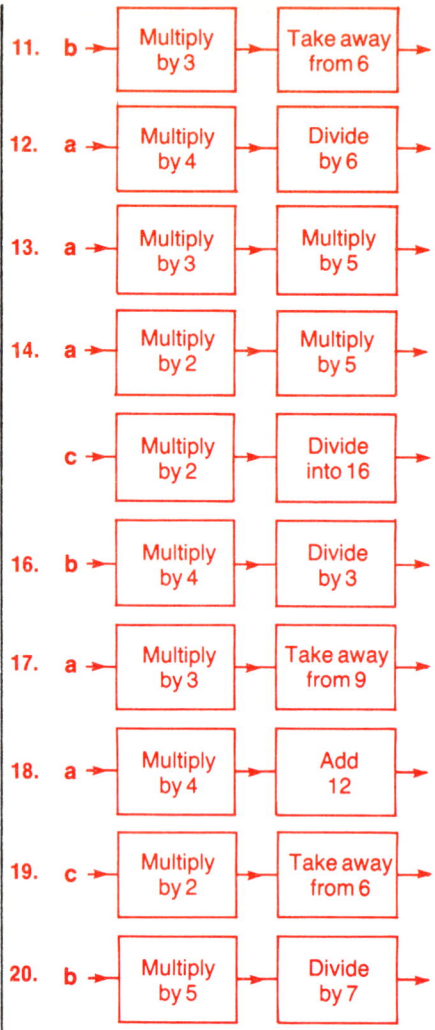

11. **b** → | Multiply by 3 | → | Take away from 6 | →

12. **a** → | Multiply by 4 | → | Divide by 6 | →

13. **a** → | Multiply by 3 | → | Multiply by 5 | →

14. **a** → | Multiply by 2 | → | Multiply by 5 | →

15. **c** → | Multiply by 2 | → | Divide into 16 | →

16. **b** → | Multiply by 4 | → | Divide by 3 | →

17. **a** → | Multiply by 3 | → | Take away from 9 | →

18. **a** → | Multiply by 4 | → | Add 12 | →

19. **c** → | Multiply by 2 | → | Take away from 6 | →

20. **b** → | Multiply by 5 | → | Divide by 7 | →

Exercise 5

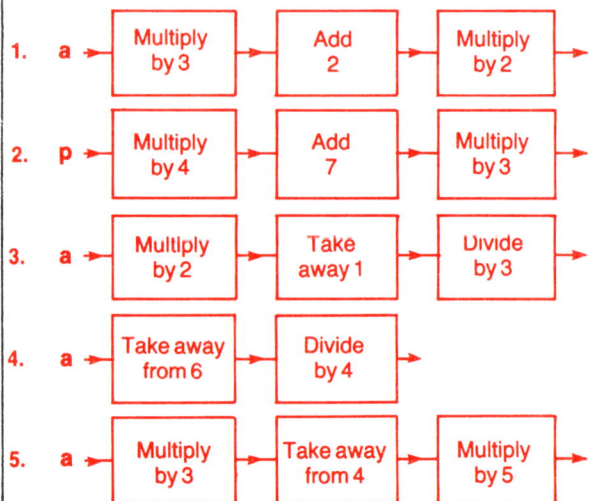

1. **a** → | Multiply by 3 | → | Add 2 | → | Multiply by 2 | →

2. **p** → | Multiply by 4 | → | Add 7 | → | Multiply by 3 | →

3. **a** → | Multiply by 2 | → | Take away 1 | → | Divide by 3 | →

4. **a** → | Take away from 6 | → | Divide by 4 | →

5. **a** → | Multiply by 3 | → | Take away from 4 | → | Multiply by 5 | →

188

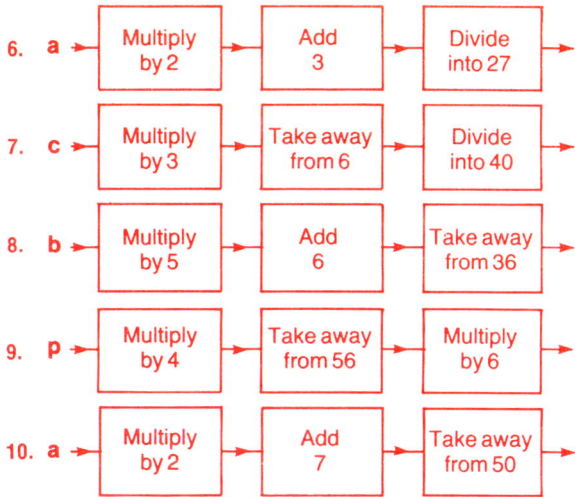

6. a → Multiply by 2 → Add 3 → Divide into 27

7. c → Multiply by 3 → Take away from 6 → Divide into 40

8. b → Multiply by 5 → Add 6 → Take away from 36

9. p → Multiply by 4 → Take away from 56 → Multiply by 6

10. a → Multiply by 2 → Add 7 → Take away from 50

2.

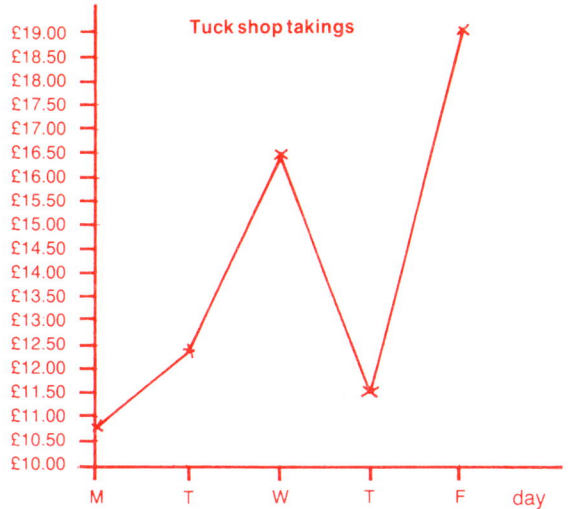

Tuck shop takings

Floor areas (pages 146-147)
Exercise 1
1. 14 m²   5. 11 m²   8. 27 m²
3. 19 m²   6. 74 m²   9. 13 m²
4. 16 m²   7. 34 m²   10. 30 m²

Averages (pages 149-150)
Exercise 1
1. 30   2. 34   3. 36   4. 60   5. 38
Exercise 2
1. £1.75   5. 60p   9. 50 g   13. £0.75
2. £4.25   6. 36p   10. 33 kg   14. 45 miles
3. £3.40   7. 301   11. 59
4. £3   8. 47   12. 20

Pounds and ounces (pages 151)
Exercise 1
1.(a) 19 oz   (c) 22 oz   (e) 36 oz   (g) 30 oz
 (b) 24 oz   (d) 32 oz   (f) 12 oz   (h) 44 oz
2.(a) 1 lb 4 oz   (c) 2 lb   (e) 1 lb 14 oz   (g) 1 lb 12 oz
 (b) 1 lb 8 oz   (d) 2 lb 4 oz   (f) 2 lb 8 oz   (h) 1 lb 7 oz

Shovels (page 155)
Exercise 1
A.
1. 4 shovels   3. 40 shovels   5. 20 shovels
2. 12 shovels   4. 16 shovels   6. 32 shovels
B.
1. 3 shovels   3. 12 shovels   5. 15 shovels
2. 6 shovels   4. 30 shovels   6. 60 shovels
C.
1. 1 shovel   3. 7 shovels   5. 3 shovels
2. 2 shovels   4. 6 shovels   6. 8 shovels

Check-up 7 (page 156)
1.(a) 28 cm   (b) 22 cm

189

3.(a) $\frac{1}{5}$ **(b)** $\frac{3}{4}$ **(c)** $\frac{7}{100}$
4.(a) $\frac{3}{5}$ **(b)** $\frac{1}{4}$ **(c)** $\frac{2}{25}$

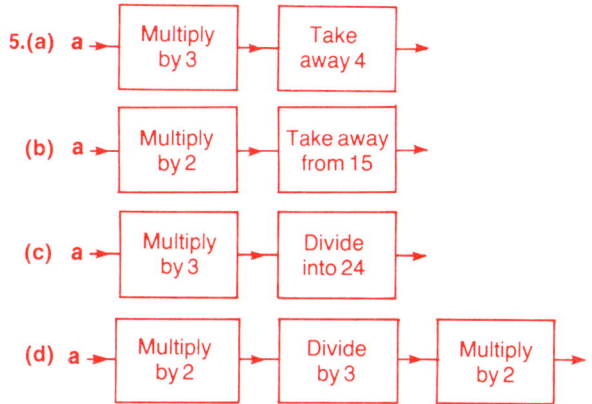

5.(a) a → Multiply by 3 → Take away 4 →

(b) a → Multiply by 2 → Take away from 15 →

(c) a → Multiply by 3 → Divide into 24 →

(d) a → Multiply by 2 → Divide by 3 → Multiply by 2 →

6.(a) 36 m²   **(b)** 36 cm²   **7.** 16
8.(a) 2 shovels   **(b)** 3 shovels

Solids (page 159)
Exercise 2

drawing	name	shape of face	No. of faces	No. of edges	No. of corners
	cube	square	6	12	8
	cuboid	rectangle	6	12	8
	tetrahedron	equilateral triangle	4	6	4

Rounding (pages 160-162)
Exercise 1
Tricia 55 kg, Barjinder 57 kg, Alice 52 kg, Renu 53 kg, Eddy 52 kg, Brian 57 kg, Gulsen 53 kg, Curtis 52 kg, Guldren 60 kg, Maxine 57 kg, Sharon 55 kg

Exercise 2
1. Roy 162 cm, Darren 159 cm, Avril 161 cm, Tricia 160 cm, Barjinder 161 cm, Alice 159 cm, Renu 160 cm, Eddy 160 cm, Brian 162 cm, Gulsen 161 cm, Curtis 160 cm, Guldren 162 cm, Maxine 161 cm, Sharon 158 cm

2. 117	215	316
37	421	1002
14	555	254

Exercise 3
40	150	380	1070
50	130	420	1350
70	100	590	2140
20	110	410	4720

Exercise 4
200	600	800	700
500	300	1200	900
300	500	4900	200
400	800	5600	1800

Gallons and pints (page 163)
Exercise 1
1. 2	**4.** ¼	**7.** 2,4	**10.** 32	**13.** 20
2. 3	**5.** 1,4	**8.** 1,2	**11.** 80	**14.** 12
3. ½	**6.** 2,2	**9.** 16	**12.** 26	**15.** 10

Missing angles (pages 164-167)
Exercise 1
1. 60°	**3.** 97°	**5.** 20°	**7.** 70°	**9.** 143°
2. 115°	**4.** 145°	**6.** 88°	**8.** 95°	**10.** 88°

Exercise 2
1. 50°	**3.** 210°	**5.** 282°	**7.** 102°	**9.** 198°
2. 220°	**4.** 140°	**6.** 140°	**8.** 37°	

Flow diagrams and equations (pages 168-171)
Exercise 1
2. 4	**5.** 27	**8.** 160	**11.** 2	**14.** 4
3. 8	**6.** 5	**9.** 7	**12.** 4	**15.** 20
4. 6	**7.** 36	**10.** 7	**13.** 6	

Exercise 2
1. 1	**3.** 4	**5.** 3	**7.** 6	**9.** 3
2. 3	**4.** 1	**6.** 1	**8.** 3	**10.** 4

Exercise 3
1. 6	**5.** 2	**9.** 2	**13.** 12	**17.** 4
2. 4	**6.** 2	**10.** 4	**14.** 4	**18.** 1
3. 4	**7.** 2	**11.** 10	**15.** 3	**19.** 4
4. 12	**8.** 1	**12.** 1	**16.** 7	**20.** 2

Angles of a triangle (page 173)
Exercise 1
1. 80°	**3.** 40°	**5.** 120°	**7.** 37°	**9.** 86°
2. 40°	**4.** 100°	**6.** 58°	**8.** 45°	

Squares (pages 174-175)
Exercise 1
A.
2. 25	**4.** 36	**6.** 81	**8.** 49	**10.** 64
3. 4	**5.** 100	**7.** 144	**9.** 1	

B.
2. $4\,cm^2$	**4.** $36\,m^2$	**6.** $64\,mm^2$	**8.** $121\,cm^2$	**10.** $144\,cm^2$
3. $9\,m^2$	**5.** $25\,mm^2$	**7.** $100\,m^2$	**9.** $169\,cm^2$	

Exercise 2
A.
2. 2 m	**3.** 7 m	**4.** 5 cm	**5.** 6 cm	**6.** 9 cm	**7.** 12 m

B.
1. 3	**2.** 1	**3.** 10	**4.** 8	**5.** 13	**6.** 11

All mixed up (page 176)
Exercise 1
1. 3¾	**4.** ¹³⁄₁₅	**7.** ¹⁴⁄₁₅ = 2⅘	**10.** ⁹⁄₁₀
2. 1⁵⁄₁₂	**5.** 3⅜	**8.** ¹⁰⁄₁₁	**11.** 3⅛
3. 3⅝	**6.** 1⁴⁹⁄₅₀	**9.** 3⅞	**12.** 1⁵⁄₉

Check-up 8 (page 177)
1.(a) 15	**(b)** 27	**(c)** 103	
2.(a) 160, 200	**(b)** 220, 200	**(c)** 480, 500	
3.(a) 1	**(b)** 19	**4.(a)** 132°	**(b)** 175°
5.(a) 3	**(b)** 4	**6.(a)** 25°	**(b)** 123°
7.(a) 64	**(b)** 36	**(c)** 16	
8.(a) 2 m	**(b)** 6 m	**(c)** 5 cm	
9.(a) 5¾	**(b)** 2½	**(c)** 4⁸⁄₂₅	

Index/Glossary

Use the index to look up where topics are in the book.
The notes are there to jog your memory!

30 PB

388 P